Anonymous

Pascal Iron Works

Vol. 1

Anonymous

Pascal Iron Works
Vol. 1

ISBN/EAN: 9783337182762

Printed in Europe, USA, Canada, Australia, Japan

Cover: Foto ©berggeist007 / pixelio.de

More available books at **www.hansebooks.com**

PRICE LIST

FOR

MORRIS, TASKER & CO.'S ILLUSTRATED CATALOGUE.

FOURTH EDITION. 1861.

Persons in ordering will please state the "Edition" they order from.

LAP WELDED AMERICAN CHARCOAL IRON BOILER FLUES.

No.	Page.		
1	1	With Plain Ends.	*For Prices, see Table below.*
2	"	With Safe Ends.	
3	"	With Safe Ends 1-16 inch larger than rest of the Flue.	
4	"	With Safe Ends, one of which is 1-16 inch larger than rest of the Flue.	
5	"	With Safe Ends, one of which is 1-4 inch less than rest of the Flue.	
6	"	Same as No. 5, with Round Shoulder.	
7	"	With Screw on one end, Collar on other.	
8	"	Cross Flues for Fire Box.	

TABLE OF STANDARD DIMENSIONS AND PRICES.

Cut to Specific Lengths to suit Purchasers.

External Diameter.	Price, per foot.	PRICE, Each Safe End, Collar or Shoulder Extra.	PRICE of Screw on one end, with Collar on the other.	PRICE of Cross Tube Each Tube Extra.	Standard Thickness.	*Nearest Wire Gauge Thickness.	Internal Diameter.	Internal Circumference.	External Circumference.	†Length of Pipe per square foot of inside surface.	†Length of Pipe per square foot of outside surface.	Internal area.	External area.	Weight, per foot.
Inches.	$ c.	$ c.	$ c.	$ c.	Inches.	Inches.	Inches.	Inches.	Inches.	Feet.	Feet.	Inches.	Inches.	lbs.
1	.15	.15	.49		0.088	13½	0.825	2.591	3.142	4.631	3.819	0.535	0.785	0.815
1¼	.18	.16	.53	1.45	0.088	13½	1.075	3.376	3.927	3.555	3.056	0.908	1.227	1.075
1½	.21	.16	.52	1.75	0.088	13½	1.325	4.162	4.712	2.883	2.547	1.379	1.767	1.311
1¾	.24	.18	.56		0.098	13	1.554	4.882	5.498	2.458	2.183	1.897	2.405	1.712
2	.28	.18	.56	2.50	0.098	13	1.804	5.667	6.283	2.118	1.909	2.556	3.142	1.991
2¼	.31	.20	.70	2.75	0.098	13	2.054	6.454	7.069	1.859	1.698	3.314	3.976	2.238
2½	.33	.22	.86	3.00	0.109	12	2.282	7.172	7.854	1.673	1.528	4.094	4.909	2.763
2¾	.38	.26	1.52		0.109	12	2.532	7.957	8.639	1.508	1.389	5.035	5.940	3.045
3	.41	.29	1.98	4.00	0.109	12	2.782	8.743	9.425	1.373	1.273	6.083	7.069	3.333
3¼	.46	.32			0.119	11	3.012	9.462	10.210	1.268	1.175	7.125	8.296	3.958
3½	.50	.35			0.119	11	3.262	10.248	10.995	1.171	1.091	8.357	9.621	4.272
3¾	.70	.38			0.119	11	3.512	11.033	11.781	1.088	1.019	9.687	11.045	4.599
4	.75	.40			0.130	10	3.741	11.753	12.566	1.023	0.955	10.992	12.566	5.320
4½	1.00	.43	Made to order.	Made to order.	0.130	10	4.241	13.323	14.137	0.901	0.849	14.126	15.904	6.110
5	1.15	.45			0.140	9½	4.72	14.818	15.708	0.809	0.764	17.497	19.635	7.220
6	1.85	.50			0.151	9	5.699	17.904	18.849	0.670	0.637	25.509	28.274	9.346
7	2.30	.60			0.172	7½	6.657	20.914	21.991	0.574	0.545	34.805	38.484	12.435
8	2.35	.75			0.182	7	7.636	23.989	25.132	0.500	0.478	45.795	50.265	16.212
9	3.50				0.193	6½	8.615	27.055	28.274	0.443	0.424	58.291	63.617	18.002
10	4.25				0.214	5½	9.573	30.074	31.416	0.399	0.382	71.975	78.540	22.19

* The thickness of Flue can be varied to order, at prices dependent upon the thickness and number of feet wanted.

† In estimating the effective steam-heating or boiler surface of Tubes or Flues; the surface in contact with air or gases of combustion (whether internal or external to the Tubes or Flues) is to be taken.

For heating liquids by steam, superheating steam, or transferring heat from one liquid or one gas to another, the mean surface of the Tubes or Flues is to be taken.

CLASS FIRST.

WROUGHT IRON WELDED TUBE, AND FITTINGS FOR SAME, &c., &c.

No. Page

9 3 Hydraulic Tube. Any diameter or thickness made to order.

10 ⎫
11 ⎭ Extra Strong and Double Extra Strong Tube.

TABLE OF STANDARD DIMENSIONS AND PRICES.

Nominal Diameter.	Actual Outside Diameter, Inches.	Thickness, Extra Strong.	Thickness, Double Extra Strong.	Actual Inside Diameter, Extra Strong.	Actual Inside Diameter, Double Extra Strong.	Price per Foot, Extra Strong.	Price per Foot, Double Extra Strong.
Inches.	Inches.	Inches.	Inches.	Inches.	Inches.	$ c.	$ c.
⅛	0.405	0.101		0.203		.10	
¼	0.54	0.12		0.32		.12	
⅜	0.675	0.129		0.431		.15	
½	0.84	0.15	0.30	0.54	0.32	.21	.42
¾	1.05	0.157	0.266	0.734	0.478	.26	.52
1	1.315	0.18	0.32	0.955	0.675	.35	.70
1¼	1.66	0.183	0.366	1.29	0.928	.52	1.05
1½	1.9	0.192	0.396	1.501	1.109	.65	1.30
2	2.375	0.229	0.458	1.917	1.459	.90	1.80
2½	2.875	0.262	0.524	2.331	1.807	1.25	2.50

12 3

WROUGHT IRON WELDED TUBE,

In Random Lengths for Steam, Gas or Water. All proved by Hydraulic Pressure of 300 pounds per square inch.

TABLE OF STANDARD DIMENSIONS AND PRICES.

Nominal Diameter.	Price per Foot.	Price per Foot, Galvanized.	Actual Outside Diameter.	Thickness.	Actual Inside Diameter.	Internal Circumference.	External Circumference.	Length of Pipe per square foot of Inside Surface.	Length of Pipe per square foot of Outside Surface.	Internal Area.	External Area.	Length of Pipe containing one cubic foot.	Weight per foot of length.	Number of threads per inch of screw.
Inches.	$ c.	$ c.	Inches.	Inches.	Inches.	Inches.	Inches.	Feet.	Feet.	Inches.	Inches.	Feet.	lbs.	
⅛	.07	.14	0.405	0.068	0.270	0.848	1.272	14.15	9.44	0.0573	0.129	2500.	0.243	27
¼	.08	[illegible]	0.54	0.088	0.364	1.144	1.696	10.50	7.075	0.1041	0.229	1385.	0.422	18
⅜	.09½	.17	0.675	0.091	0.494	1.552	2.121	[illegible]	[illegible]	[illegible]	[illegible]	[illegible]	[illegible]	18
½	.10½	.18	0.84	0.109	0.622	1.954	2.652	6.13	4.502	0.3048	0.554	472.4	0.845	14
¾	.13	.22	1.05	0.113	0.824	2.589	3.299	4.635	3.637	0.5333	0.866	270.	1.126	14
1	.17½	.30	1.315	0.134	1.049	3.296	4.134	3.679	2.903	0.8627	1.357	166.9	1.670	11½
1¼	.23	.40	1.66	0.140	1.380	4.335	5.215	2.768	2.301	1.496	2.164	96.25	2.258	11½
1½	.32	.50	1.9	0.145	1.611	5.061	5.969	2.371	2.01	2.038	2.835	70.66	2.694	11½
2	.40	.70	2.375	0.154	2.067	6.494	7.461	1.848	1.611	3.355	4.430	42.36	3.631	11½
2½	.60	1.20	2.875	0.204	2.468	7.754	9.032	1.547	1.328	4.783	6.491	30.11	5.733	8
3	1.24	1.75	3.5	0.217	3.067	9.636	10.996	1.245	1.091	7.388	9.621	19.49	7.541	8
3½	1.60	2.25	4.0	0.226	3.548	11.146	12.566	1.077	0.955	9.887	12.566	14.56	9.055	8
4	1.80	2.75	4.5	0.237	4.026	12.648	14.137	0.949	0.849	12.730	15.904	11.31	10.728	8
4½	2.40	3.25	5	0.247	4.506	14.156	15.708	0.848	0.765	15.939	19.635	9.03	12.487	8
5	2.75	3.75	5.563	0.259	5.045	15.849	17.475	0.757	0.620	19.990	24.299	7.20	14.502	8
6	4.00	3.75	6.625	0.280	6.065	19.054	20.813	0.63	0.577	28.889	34.471	4.98	18.767	8
7	5.50	[illegible]	7.625	0.301	7.023	22.063	23.954	0.544	0.505	38.737	45.663	3.72	23.410	8
8	7.00	9.25	8.625	0.322	7.982	25.076	27.096	0.478	0.444	50.039	58.426	2.88	28.348	8
9			9.688	0.344	9.001	28.277	30.433	0.425	0.394	63.633	73.715	2.26	34.077	8
10			10.75	0.366	10.019	31.475	33.772	0.381	0.355	78.838	90.762	1.80	40.641	8

Taper of threads, 1 to 32 on each side.

CLASS FIRST.—*Continued.*

No.	Page.	Nominal Diameter. Inches.	⅛	¼	⅜	½	¾	1	1¼	1½	2	2½	3	3½	4	4½	5	6	7	8	Price for each.
			$ c.	$ c.	$ c.	$ c.	$ c.	$ c.	$ c.	$ c.	$ c.	$ c.	$ c.	$ c.	$ c.	$ c.	$ c.	$ c.	$ c.	$ c.	
13	2	Long Screws,		.20	.24	.26	.35	.50	.70	.92	1.45	2.26	3.50	4.60							"
14	"	Wrought Bends,	.16	.18	.20	.23	.30	.40	.55	.76	1.05										"
15	"	Spring from the Main, 16°,																			
16	"	" " " 8°,	.11	.13	.13	.15	.18	.23	.34	.46	.63										"
17	"	" " " 17°,																			
18	"	Sockets,	.05	.06	.07	.08½	.11	.16	.22	.30	.46	.70	1.10	1.50	2.00	2.60	3.20	4.70	6.40	8.25	"
19	"	Reducing Sockets, and Right and Left Hand Sockets,	.06	.07	.08	.09½	.12	.15	.23	.31	.50	.77	1.20	1.65	2.20	2.60	3.18	5.28	7.00	9.60	"
20	"	Caps,	.05	.06	.07	.08½	.11	.16													"
21	"	Plugs,	.05	.06	.07	.08½	.11	.16													"
22	"	Lock Nuts,	.05	.06	.07	.08½	.11	.16													"
23	"	Close Nipples,	.05	.06	.07	.08½	.11	.16													"
24	"	Lock Nuts,	.05	.06	.07	.08½	.11	.16													"
25	"	Bushings,	.05	.06	.07	.08½	.11	.16													"
26	"	Shoulder Nipples,	.05	.06	.07	.08½	.11	.16													"
(27	"	Gas Hooks, see Class Fourth,) &c.																			"
28	"	Crosses,	.13	.15	.18	.22	.28	.41	.55	.80	1.30	1.85	2.80	3.90	5.00	6.50	8.20	11.90	16.20	21.60	"
28½	"	Globe Crosses, Reducing and Corner Fittings. Same price as Crosses, Tees, or Elbows of usual forms; the largest outlet being taken as standard for price.																			
29	"	Tees,	.10	.11	.14	.17	.22	.32	.45	.61	.92	1.42	2.20	2.85	4.00	5.15	6.25	9.30	12.75	16.52	"
30	"	Elbows,	.07	.08½	.10	.12	.15	.22	.34	.47	.66	1.00	1.66	2.39	3.00	4.15	4.50	6.80	9.30	12.00	"
31	"	Return Bends, Close Pattern. Distance from centre to centre, inches, 1⅝ 1¾																			
32	"	" " Wide " " " " " 2 2½																			
33	1	" " Square "			.12c	.14c	.22c	.30	.50	.70	1.00	1.50									"
36	"	Collars.—Made to order.																			

34 "

MANIFOLDS.

All Manifolds have left-hand Taper Screws, unless especially ordered, either with running or right-hand Taper Screws.

They are also open at both ends, with right-hand Taper Thread of same size as Outlets, irrespective of size of body. Either or both ends, however, can be drilled out to take Screws the size of body, or any size less than the size selected as size of body.

Back or Side Outlets can be attached to any manifold pattern to order; but in ordering, care must be taken to describe the place of attachment of Back or Side Outlets.

Back or Side Outlets of same size as Front Outlets, will be charged as additional Front Outlets; other sizes, proportionally.

Number of Outlets.		2	3	4	5	6	7	8	10	12	14	16	18	20
	Size of Body.	$ c.	$ c.	$ c.	$ c.	$ c.	$ c.	$ c.	$ c.	$ c.	$ c.	$ c.	$ c.	$ c.
For 1 inch Pipe,	1¼	.62	.88	1.08	1.28	1.48	1.68	1.88						
	1½	.74	.99	1.21	1.43	1.66	1.88	2.11						
	2	.85	1.10	1.35	1.60	1.85	2.10	2.35	2.85	3.35	3.85	4.35	4.85	5.35
For ¾ inch Pipe.	1	.43	.55	.68	.80	.93	1.05	1.18						
	1¼	.49	.64	.78	.93	1.07	1.22	1.36						
	1½	.56	.72	.89	1.05	1.22	1.38	1.55	1.88	2.21	2.54	2.87	3.20	3.53

Manifolds, for ½ inch Pipe Outlets, body size of 1 inch pipe, at 18 cents each, plus 12 cents for each outlet.
" " 1¼ " " " " 2 " " 45 " " " 30 " " "
" " 1½ " " " " 2½ " " 63 " " " 42 " " "
" " 2 " " " " 3 " " 93 " " " 63 " " "
" " 1 " " are 2½ from Centre to Centre.
" " ¾ " " " 2 " " "

CLASS FIRST.—*Continued.*

No. Page [illegible]

MANIFOLD VALVES.

Number of Outlets		2	3	4	5	6	7	8	10
	Size of Body	\$ c.	\$ c.	\$ c.	\$ c.	\$ c.	\$ c.	\$ c.	\$ c.
Manifold Valves, For 1 in. pipe	2	4.25	6.50	8.00	9.75	11.75	13.75	15.50	19.25
" " ¾ "	1½	3.25	4.75	6.25	7.75	9.25	10.75	12.25	15.25

No. Page [illegible]

COIL STANDS.

Number of Pipes High.		2	4	6	8	10	12	14	16	18	20	22	24
	Size of Pipe	\$ c.	\$ c.	\$ c.	\$ c.	\$ c.	\$ c.	\$ c.	\$ c.	\$ c.	\$ c.	\$ c.	\$ c.
Coil Stands, per pair,	1	.44	.62	.90	.98	1.16	1.28	1.98	2.22	2.46	2.70	2.94	3.18
	¾	.34	.48	.62	.76	.90	1.04	1.18					

No. Page [illegible]

CAST IRON FLANGES.

Size of Pipe upon which they will screw.	¼	⅜	½	¾	1	1¼	1½	2	2½	3	3½	4	Inches
Diameter of Flanges, Inches.	\$ c.	\$ c.	\$ c.	\$ c.	\$ c.	\$ c.	\$ c.	\$ c.	\$ c.	\$ c.	\$ c.	\$ c.	each.
12									2.37	2.36	2.53	2.79	"
11								1.80	1.97	2.09	2.20	2.18	"
10								1.60	1.69	1.78	1.94	2.04	"
9½							1.55	1.41	1.53	1.62	1.75	1.89	"
9						1.19	1.24	1.39	1.39	1.50	1.62	1.75	"
8½					1.00	1.05	1.10	1.25	1.25	1.38	1.57		"
8					.79	.84	.90	1.07	1.17	1.27	1.40		"
7½				.76	.80	.84	.88	.97	1.08	1.11			"
7				.67	.71	.75	.79	.89	.99				"
6½			.55	.58	.62	.67	.72	.80	.90				"
6		.46	.48	.51	.55	.59	.64	.73					"
5½		.39	.42	.45	.49	.53	.58						"
5	.32	.34	.36	.39	.44	.48							"
4½	.28	.30	.32	.34	.38	.42							"
4	.24	.25	.27	.30	.34								"

No. Page [illegible]

CAST IRON FLANGES WITH BOLTS, WITH DIAMETERS CORRESPONDING TO THOSE OF FLANGES UPON CAST IRON PIPE, FOR CONNECTING WROUGHT AND CAST IRON PIPE TOGETHER.

Diameter of Flanges.	Number of Bolts.	Size of Bolts.	Size of Pipe upon which they will Screw.								
			3	3½	4	4½	5	6	7	8	Inches
Inches.		Inches.	\$ c.	\$ c.	\$ c.	\$ c.	\$ c.	\$ c.	\$ c.	\$ c.	
6½	4	½	1.05								
7	4	5-16	1.20	1.30							
7½	4	5-16	1.30	1.40							
8	4	5-16		1.65	1.75						
8½	5	⅜			1.90	2.05					
9	5	⅜			2.05	2.20					
9½	5	⅜				2.50	2.55				
10	6	⅜					2.75				
10½	6	⅜					2.90	3.30			
11	7	⅜						3.60			
12	8	⅜						4.00	4.60		
13½	8	⅜							4.80	5.50	
14	10	⅜								6.50	

CLASS FIRST—*Continued.*

TAPPING FLANGES, TO ORDER.

Sizes	¼	⅜	½	¾	1	1¼	1½	2	2½	3	3½	4
Price, net	7 cts.	8 cts.	9 cts.	10 cts.	14 cts.	17 cts.	21 cts.	25 cts.	[illegible]	[illegible]	[illegible]	$1.00

No.	Page												
139	13	Hook Plates,	Number of Hooks,	1	2	3	4	5	6	7	8	10	12
142	13	Single Hook Plates,	For ¾ inch Pipe, price,	6 cts.	10 cts.	14 cts.	18 cts.	22 cts.	26 cts.	30 cts.	38 cts.		
143	13	Single Hooks,	For 1 inch Pipe, price,	8 "	13 "	16 "	23 "	28 "	33 "	38 "	43 "		
140	13	Corner Plates,	For ¾ inch Pipe, price,	8 "	12 "	18 "	23 "	28 "	33 "	38 "	43 "		
140 a	13	Ring Plates,	For 1 inch Pipe, price,	10 "	16 "	22 "	28 "	34 "	40 "	46 "	52 "		

141	13	Rosette Plates,	Number of Pipes,	2	4	6	8	10	12
			For ¾ inch Pipe, price,	24 cts.	40 cts.	56 cts.	72 cts.	88 cts.	$1.04
			For 1 inch Pipe, price,	30 "	50 "	70 "	90 "	$1.10	1.30

144 13 Stand Brackets—same price as Hook Plates.
145 " Close Pipe Supports—same price as Ring Plates.
145 a. " Wall Plates— " " "
146 " Movable Hook Plates for Laundry Coils—same price as Rosette Plates.
147 " Expansion Hanger (with one foot of rod) for ¾ inch, 40c.; 1 inch, 45c.; 1¼ inch, 50c.; 1½ inch, 55c.; 2 inch, 65c.; 2½ inch, 75c.; 3 inch, 90c.
148 " Expansion Hanger for Large Pipe—3½ inch, $1.50; 4 inch, $1.75.
149 " Triple Expansion Hanger, }
150 " Hanger in halves, } Made to order.
151 " Wall Hanger, }
159 15 Radiators for Walls, with Manifolds, }
160 " " " " " Return Bends, }
161 " " Corners, with Manifold Valves, }
162 " " Drying Closets, with Return Bends, }
Same price per foot of Pipe used as box coils (No. 165,) with additional charge for all Manifolds, per list of prices, and additional charge for difference of prices between Rosette or Movable Hook Plates and ordinary Hook Plates.

165 17

BOX COILS.

ALL MADE UP IN STANDS AND READY FOR USE, ANY NUMBER OF PIPES HIGH OR WIDE.

Length of Pipes in Coils,	2 ft.	2 ft. 3 in.	2 ft. 6 in.	3 ft.	3 ft. 6 in.	4 ft.	4 ft. 6 in.	5 ft.	6 ft.	7 ft.	8 ft.	9 ft.	10 ft.	12 ft.
Price per foot, 1 in. Tube,	55 cts.	51 cts.	48 cts.	43 cts.	39 cts.	37 cts.	35 cts.	32 cts.	31 cts.	29 cts.	28 cts.	27 cts.	26 cts.	25 cts.
" " ¾ "	41 "	38 "	35 "	32 "	29 "	27 "	26 "	25 "	23 "	22 "	21 "	20 "	20 "	19 "

CLASS SECOND.

IRON BODY AND BRASS FITTINGS, VALVES, WITH OR WITHOUT FLANGES.

No.	Page	Inches,	1¼	1½	2	2½	3	3½	4	4½	5	6	8	10	12	
43	5	Steam Valves, Iron Body, with Brass Fittings, Globe and Angle Flanged with outside screw; 2½ inches and below, with inside screw; with brass discs and spherical seats,					6½	7½	9	9½	9	10½	12½	15	17½	Diameter of Flanges Inches Diameter.
							9.25	10.24	11.72	12.21	13.19	15.17	19.11	23.05	27	Ins. face to face.
							4	5	5	5	6	6	8	10	10	Number of Bolts.
							½	⅝	$\frac{9}{16}$	$\frac{9}{16}$	$\frac{9}{16}$	¾	⅞	¾	⅞	Inches size of Bolts.
			$ c.	$ c.	$ c.	$ c.	$ c.	$ c.	$ c.	$ c.	$ c.	$ c.	$ c.	$ c.	$ c.	Price each.
43a	5	Same—with screwed ends,	5.00	5.75	7.50	11.00	15.00	12.50	24.00	30.00	35.00	45.00	70.00	100.00	120.00	"
43b	5	Globe, Vertical and Horizontal Check Valves, do do do	4.00	4.75	7.50	11.00	15.00									"
43c	5	Globe Safety Valves, do do do	6.50	8.00	11.50	15.50	20.00	25.00	31.00	37.00	41.00	56.00	85.00	112.50	140.00	"
43d	5	Globe Cross Valves, do do do	4.40	5.25	8.25	12.00	16.50	21.00	26.50	33.00	38.50	49.50	77.00	110.00	143.00	"
43e	"	Globe Back Pressure Valves, do do														
42	5	Transfer Valves, do do do														

FINISHED BRASS WORK.

No.	Page	Inches,	¼	⅜	½	¾	1	1¼	1½	2	2½	3	3½	4	4½	5	6	Price per each
			$ c.	$ c.	$ c.	$ c.	$ c.	$ c.	$ c.	$ c.	$ c.	$ c.	$ c.	$ c.	$ c.	$ c.	$ c.	
53	6 and 7	Patent Hydrant Cocks,		4.00	4.75													"
54	"	Basin Cocks,		1.25														"
55	"	Brass Bib Cocks, lever handle, screw tail,		1.50	1.75	2.00	3.50	4.75	6.50	8.50								"
56	"	Brass Bib Cocks, hose screw,		1.50	1.75	2.00	3.50	4.75	6.50	8.50								"
57	"	Brass Bib Cocks, plain,		1.50	1.75	2.00	3.50	4.75	6.50	8.50								"
58	"	Brass Vat Cocks—made to order,																"
60	"	Double Oil Cocks, (diameter of Globe,)									5.00		10.00			12.00		"
62	"	Try Cocks for Purifiers,		1.75														"
63	"	Oil Cups, (diameter of Bowl,)							1.50	1.65								"
64	"	Brass Hinges,										1.00	4.50					"
65	"	Mercury Cups,				.75												"
66	7	Glass Water Level Indicator,																"
66a	"	Angle Cocks for Sugar House use—made to order,																"
90	8	Bath Tub Overflow,						2.50	2.75									"
91	"	Wash Tub Waste,							2.75									"
92	"	Basin Wastes and Chains,				.40												"
93	"	Bath Tub Strainers,						2.00										"
94	"	Sink Wastes with Union,						2.50										"
95	"	Sink Wastes with Screw,						2.00										"
86	"	Malleable Iron Unions,	.25	.35	.40	.55	.75	1.00	1.35	1.75	2.50	3.50	4.50					"
87	"	Brass Unions, inside screw,	.45	.65	.85	1.00	1.40	2.00	2.25	3.25	4.25							"
88	"	Bath Boiler Unions, straight,	.45	.65	.85	1.00	1.40	2.00	2.25	3.25	4.75							"
89	"	Brass Unions, inside screw,	.45	.65	.85	1.00	1.10	2.00	2.25	3.25	4.75							"
90	"	Bath Boiler Unions, (Gooseneck,)				1.00	1.50											"
91	"	Boiler Ferrule,			.25	.45	.60											"
92	"	Hose Couplings for Pavements,				.50	.75	1.12	1.75									"
93	"	Brass Steam Whistles,										11.00		14.00		18.00	25.00	"
94	"	Fire Hose Couplings,						2.00	2.75	3.00	4.00	4.50	5.00	6.00				"
95	"	Coupling Wrenches,—made to order,																"
96	"	Branch Pipes,—made to order,																"
98	9	Differential Pressure Valves for Reducing the Pressure of Steam in Pipes to any requisite pressure, } Made to order																"
99	"	Combined Stop and Safety Valve,										10.00						"
100	"	Copper Strainers,				3.00	3.38	4.12	5.00	6.25								"
101	"	Bath Tub Overflow or Stand Pipe, with Unions,						2.10	2.35									"

CLASS THIRD.

BRASS VALVES, &c., &c.

No.	Page	Inches,	⅛	¼	⅜	½	¾	1	1¼	1½	2	2½	3	3½	4	Price per
			$ c.	$ c.	$ c.	$ c.	$ c.	$ c.	$ c.	$ c.	$ c.	$ c.	$ c.	$ c.	$ c.	piece.
44	5a	Brass Globe and Angle Steam Valves; Loose Discs and spherical seats; above ½ of an inch; screwed ends,		1.15	1.25	1.65	2.40	3.45	4.15	6.00	9.50	15.00	20.00			"
44b	5a	Brass Globes, Safety Valves, do do do				3.00	4.00	5.50								"
45	5a	Brass Angle Valves, do do do		1.15	1.32	1.65	2.40	3.45	4.15	6.00	9.50	15.00	20.00			"
46	5a	Brass Cross Valves, do do do		1.27	1.40	1.82	2.64	3.90	5.25	6.60	10.50	16.50	22.00			"
47	5a	Brass Bath-tub Valves, do do do					2.00									"
48	5a	Brass Bath-tub Valves, double, do do do					3.40									"
46a	5a	Brass Corner Valves, do do do														"
49	5a	Brass Angle Bib Valves, do do do				2.00	3.00									"
45a	5a	Brass Angle Pressure Valves, do do do														"
50 / 50a	5a / 5a	Brass Horizontal and Vertical Globe Check Valves, do do do		1.05	1.25	1.50	2.20	3.10	4.30	5.80	8.65	12.30	18.00			"
51	5a	Brass Double Disc Governor Valves, do do														"
51a	5a	Brass Double Disc Pump Valves, do do do														"
37	4	Lamp Cocks,		.75												"
38	"	Lamp Tube,					2.00									"
39	"	Brass Cocks for Steam or Water,			1.00	1.50	2.20	3.00	4.25	6.00	10.50	16.50	24.00	33.00	48.50	"
39a	"	Iron Cocks " " with Brass Plugs,				1.35	1.75	2.50	3.00	5.00	8.75	13.75	25.00			"
40	"	Iron Cocks, with or without Flanges,			.80	1.00	1.40	2.00	2.50	3.50	5.50	8.50	12.50	17.50	23.50	"
41	"	Three-way Iron Cocks,														"
39b	7	Brass Cocks with Soldering Unions,			1.30	1.90	2.70	3.60	5.00	7.80	12.25	19.00	27.75	38.00	58.00	"
42a	5a	Brass Transfer Valves,														"
59	6	Brass Swing Joints,			1.25	1.50	1.75	3.25	4.25	5.50						"
61	"	Brass Expansion Joints,				1.75	2.00	3.00	3.75	4.25	6.50	12.50	15.00	17.00	22.00	"
67	"	Brass Vacuum Valves,				1.25	1.75	2.50	3.50							"
68	"	Brass Gauge Valves, loose seats, stuffing boxes with glands, Whole length over all from end of screw to wheel, 6 8 9½ inches. $2.50 $3.25 $4.75 Price each.														
69	"	Brass Gauge Cocks, loose seats, whole length, 5 inches, Price, $1.50 each.														
70	"	Brass Gauge Cocks, whole length, 4 inches, price $1.25 "														
71	"	Brass Gauge Cocks, do do 5 do do 1.50 "														
72	"	Air or Pet Cocks,	.90	1.00	1.30											"
74 / 75	" / "	Brass Gas Meter and Service Cocks, with Male or Female Screws,			8.60	10.50	15.00	12.50	40.00	56.50	160.00					pairs.
76	"	do do do with Unions,			10.50	13.50	14.00	27.00	47.00	66.00	116.00					"
77	"	Iron Pipe Stops, all brass,					1.50	2.20	3.75	4.75	7.60					each.
78	"	Brass Crooked Hydrant Cocks and Waste,				1.00	1.50	2.13								"
79	"	Brass Straight Hydrant Cocks and Waste,				1.00	1.50	2.15								"

CLASS FOURTH.

COILS, COIL SCREENS, CAST IRON PIPES, &c. &c.

No.	Page		
151	14	Taper Coils,	Any size or pattern made to order.
152	"	Snap Coils,	
154	"	Heater Coils,	
154a	"	Heater Coils,	
155	"	Steam Gauge Coils,	
156	"	Tymp Coils,	
157	"	Flat Coils for Tanks,	
158	"	Super Heating Steam Coils,	
163	16	Ornamental Radiators for Walls,	$35 00
164	"	Cast Iron Pedestal Radiators—three row,	[illegible] 00
		" " "—two row,	85 00
166	17	Screens for Covering Box Coils.	
167	17a	" " "	
168	17a	" " "	
169	17b	" " "	
169a	17b	Screens for Recess Radiator.	

COIL SCREENS, WITH MARBLE TOP. FOUR SIDES. Size in the clear, 60 inches long, 18¾ inches wide, 31½ inches high. Gold Bronzed, $45. Plain Bronzed, $41.60.

No. 2 SCREENS, WITH MARBLE TOPS. FOUR SIDES.

Size in the clear.	Gold Bronzed.	Plain Bronzed.	Size in clear.	Gold Bronzed.	Plain Bronzed.
66¾ in. long, 18¾ in. wide, 32 in. high,	$57.00	$52.75	47¾ in. long, 18¾ in. wide, 31 in. high,	$45.00	[illegible]
" " 16¾ " " "	55.75	51.60	" " 16¾ " " "	41.75	38.65
" " 14¾ " " "	54.50	50.45	" " 14¾ " " "	40.50	37.50
" " 12¾ " " "	53.25	49.30	" " 12¾ " " "	39.25	36.35
" " 10¾ " " "	52.00	48.15	" " 10¾ " " "	38.00	35.20
" " 8¾ " " "	50.75	47.00	" " 8¾ " " "	36.75	34.05
60¾ " 18¾ " " "	53.00	49.05	42¾ " 18¾ " " "	41.00	37.95
" " 16¾ " " "	51.75	47.90	" " 16¾ " " "	39.75	36.80
" " 14¾ " " "	50.50	46.75	" " 14¾ " " "	38.50	35.65
" " 12¾ " " "	49.25	45.60	" " 12¾ " " "	37.25	34.50
" " 10¾ " " "	48.00	44.45	" " 10¾ " " "	36.00	33.35
" " 8¾ " " "	46.75	43.30	" " 8¾ " " "	34.75	32.20
54¾ " 18¾ " " "	49.00	45.35	36¾ " 18¾ " " "	35.00	30.05
" " 16¾ " " "	47.75	44.20	" " 16¾ " " "	37.75	34.90
" " 14¾ " " "	46.50	43.05	" " 14¾ " " "	36.50	33.75
" " 12¾ " " "	45.25	41.90	" " 12¾ " " "	35.25	32.60
" " 10¾ " " "	44.00	40.75	" " 10¾ " " "	34.00	31.45
" " 8¾ " " "	42.75	39.60	" " 8¾ " " "	32.75	30.30
48¾ " 18¾ " " "	45.00	41.65	30¾ " 18¾ " " "	37.00	34.25
" " 16¾ " " "	43.75	40.50	" " 16¾ " " "	35.75	33.10
" " 14¾ " " "	42.50	39.35	" " 14¾ " " "	34.50	31.95
" " 12¾ " " "	41.25	38.20	" " 12¾ " " "	33.25	30.80
" " 10¾ " " "	40.00	37.05	" " 10¾ " " "	32.00	29.65
" " 8¾ " " "	38.75	35.90	" " 8¾ " " "	30.75	28.50

CLASS FOURTH—*Continued.*

NO. 2 SCREENS, WITH MARBLE TOPS. FOUR SIDES.

SIZE IN THE CLEAR.	GOLD BRONZED.	PLAIN BRONZED.	SIZE IN THE CLEAR.	GOLD BRONZED.	PLAIN BRONZED.
55¾ in. long, 18¾ in. wide, 33¾ in. high.	$48.60	$43.60	52¾ in. long, 18¾ in. wide, 32¾ in. high.	$44.70	$40.10
" " 16¾ " " "	46.70	42.60	" " 16¾ " " "	43.60	39.00
" " 14¾ " " "	45.80	41.60	" " 14¾ " " "	42.50	38.00
" " 12¾ " " "	44.70	40.60	" " 12¾ " " "	41.40	37.00
" " 10¾ " " "	43.00	39.60	" " 10¾ " " "	40.30	36.00
" " 8¾ " " "	42.50	38.60	" " 8¾ " " "	39.20	35.00
54¾ " 18¾ " 32¾ "	41.10	37.60	42¾ " 18¾ " 31¾ "	35.00	31.90
" " 16¾ " " "	40.20	36.56	" " 16¾ " " "	33.90	30.80
" " 14¾ " " "	39.10	35.50	" " 14¾ " " "	32.80	29.80
" " 12¾ " " "	37.90	34.50	" " 12¾ " " "	31.70	28.80
" " 10¾ " " "	36.50	33.00	" " 10¾ " " "	30.60	27.80
" " 8¾ " " "	35.80	32.50	" " 8¾ " " "	29.50	26.80
48¾ " 18¾ " " "	38.20	34.70	39¾ " 18¾ " " "	33.20	30.25
" " 16¾ " " "	37.10	33.70	" " 16¾ " " "	32.20	29.25
" " 14¾ " " "	36.00	32.50	" " 14¾ " " "	31.20	28.25
" " 12¾ " " "	34.90	31.70	" " 12¾ " " "	30.20	27.25
" " 10¾ " " "	33.60	30.70	" " 10¾ " " "	28.30	26.25
" " 8¾ " " "	32.70	29.70	" " 8¾ " " "	27.20	25.25
45¾ " 18¾ " 31¾ "	35.60	33.25	36¾ " 16¾ " " "	32.00	29.10
" " 16¾ " " "	34.50	32.25	" " 16¾ " " "	30.90	28.10
" " 14¾ " " "	33.40	31.25	" " 14¾ " " "	29.80	27.10
" " 12¾ " " "	33.30	30.25	" " 12¾ " " "	28.70	26.10
" " 10¾ " " "	32.30	29.25	" " 10¾ " " "	27.60	25.10
" " 8¾ " " "	31.40	28.25	" " 8¾ " " "	26.50	24.10

Screens of THREE SIDES, and any other size than the above, made to order. Ten per cent. to be deducted from these prices for Three Sided Screens.

WHEEL REGISTERS.

☞ *Customers ordering, will please use the description given on the List in describing the kinds wanted.*

SIZE OF OPENING.	No. 1, BEST BLACK.	*VENTILATORS, WITH PULLEY AND BOX CORDS.	WHITE ENAMELLED.	FANCY ENAMELLED.	GOLD BRONZED.	LEAF PLATED.	IRON BORDER FRAMES. BRONZED.	IRON BORDER FRAMES. BLACK.
4½ by 6½ inches, each	$1.20	$1.40	$2.50	$3.50	$3.30	$5.75		
4 by 10 " "	1.50	1.70	3.30	4.40	2.00	6.50		
4 by 15 " "	2.10	2.30	4.00	5.25	2.70	8.00		
4 by 18 " "	2.25	2.45	4.50	5.50	2.85	10.00		
6 by 8 " "	1.60	2.00	3.30	4.40	2.10	7.25	$1.00	$0.75
6 by 10 " "	1.80	2.00	3.70	4.90	2.25	8.75	1.15	.84
6 by 14 " "	2.40	2.60	4.50	5.75	3.00	11.50	1.25	1.00
6 by 16 " "	2.00	3.20	5.25	6.50	3.60	13.50	1.60	1.20
8 by 18 " "	3.16	3.75	6.50	7.25	4.15	16.25	1.85	1.36
12 by 12 " "	3.00	4.00	6.50	7.75	4.50	16.25	1.60	1.30
10 by 10 " "	2.80	3.00	5.00	6.25	3.40	12.00	1.45	1.10
8 by 8 " "	2.00	2.20	4.00	5.10	2.60	8.25	1.20	.90
8 by 10 " "	2.20	2.40	4.20	5.40	2.75	10.00	1.30	1.00
8 by 12 " "	2.50	2.70	4.60	5.80	3.10	11.75	1.40	1.10
9 by 14 " "	3.25	3.45	5.50	6.75	3.90	12.80	1.60	1.20
10 by 14 " "	3.75	3.95	6.00	7.30	4.40	17.00	1.80	1.40
10½ by 10½ " "	4.40	4.60	6.75	8.25	5.10	20.00	2.00	1.55
12 by 15 " "	5.25	5.50	7.75	9.25	6.20	26.50	2.25	1.80
14 by 22 " "	7.25	7.50	10.00	12.00	8.25	.	2.80	2.20
15 by 25 " "	8.20	8.75	13.00	15.00	10.25	.	3.50	3.00
20 by 26 " "	12.00	12.50	18.00	.	14.00	(Ship Scuttle or Sliding, 20 by 26, $3.75 Plate.)		
30 by 30 " "	25.00	(Large size, for Stores and Public Buildings.)				" " " 30 by 30, 4.50 "		

8 by 10 and 8 by 12, with heavier Tops, for Floors, same price as above.

* If Oblong Ventilators are wanted to stand endwise, persons will note the fact in ordering, otherwise they will be sent to be used the other way.

CLASS FOURTH.—*Continued.*

GRATE OR FIREPLACE REGISTERS. TOPS CIRCULAR.

SIZE OF OPENING.	No. 1, BEST BLACK.	VENTILATORS WITH FIXTURES FOR CORDS.	WHITE ENAMELED.	FANCY ENAMELLED.	GOLD BRONZED.	LEAF PLATED.
11 by 13 inches each,	$3.25	$3.50	$5.00	$6.80	$3.90	$15.00
13 by 15 " "	4.25	4.50	6.50	7.80	4.90	19.00
16 by 17 " "	5.75	6.00	8.50	9.80	6.50	26.00
8 by 14 " "	2.80	With Rear Cords	5.20	6.25	3.40	13.50
9 by 12 " "	2.75	3.00	5.20	6.20	3.30	12.00
10 by 14 " "	3.25	3.50	5.40	6.60	3.90	14.00
12 by 17 " "	4.50	4.75	7.25	8.50	5.25	22.00
8 by 14 " "	2.80	3.05	. .	. .	3.40	New Pattern Carved Top—very ornamental for Ventilators.
9 by 12 " "	2.75	3.00	. .	. .	3.30	
10 by 14 " "	3.25	3.50	. .	. .	3.90	

Raised Top Ventilators, with Fixtures for Cords, 6 by 8 White Japanned, $1.90, ... White Enameled, $3.50, ... Fancy Enameled, $4.60
" " " " " " 6 by 10 " " 2.10, ... " " 4.00, ... " " 5.20
" " " " " " 8 by 10 " " 2.50, ... " " 4.50, ... " " 5.75
" " " " " " 8 by 12 " " 2.80, ... " " 5.00, ... " " 6.25

Round Revolving Registers, 10 inches, $0.85—Soap Stone, $0.90. Net prices.
" " " 12 " .95— " " 1.20, "
" " " 14 " 1.25— " " 1.50, "
" " " 16 " 1.75— " " 2.00, "

Cast Iron Flue Dampers, for Hot Air Flues, 4 by 6 inches in the opening. Net prices, $0.24
" " " " " 4 by 8 " " " " .30
" " " " " 4 by 12 " " " " .42
" " " " " 6 by 9 " " " " .48
" " " " " 6 by 12 " " " " .54
" " " " " 9 by 9 " " " " .62

TWO FAN REGISTERS. *A New Article.*

	REGISTERS, BLACK OR WHITE.	VENTILATORS, BLACK OR WHITE.	GOLD BRONZED REGISTERS AND VENTILATORS.	EXTRA VENTILATORS, FULL GILT, TO ORDER.
5 inches, round, each,	$0.80	$0.80	$1.10	
6 " " "	1.00	1.00	1.40	
7 " " "	1.10	1.10	1.50	$4.75
8 " " "	1.40	1.40	1.80	5.00
10 " " "	2.00	2.00	2.50	5.75

Patent Lock or Asylum Registers, operated by a key, 10 by 9 inches, ... $2.80
" " " " " " 10 by 12 " . 3.00
" " " " " " 10 by 16 " 4.50
" " " " " " 10 by 20 " . . . 5.50

	REGISTERS, BLACK OR WHITE.	REGISTERS, PLATED CENTRES.	VENTILATORS, WITH FIXTURES FOR CORDS.
5 inches, round,	$0.94	$1.10	$0.94
6 " "	1.20	1.35	1.20
7 " "	1.30	1.45	1.30
8 " "	1.60	1.80	1.60
9 " "	2.00	2.20	2.00
10 " "	2.25	2.50	2.25
12 " "	2.80	3.00	2.80
4 by 8 inches, square,	1.20	1.35	1.20
5 by 8 " "	1.45	1.60	1.45
8 by 8 " "	2.00	2.70	2.00

CLASS FOURTH.—*Continued.*

EIGHT FAN REGISTERS, WITH PLATED STAR CENTRES.

	No. 1, Best Black.	White Enameled.	Fancy Enameled.	Ventilators, with Fixtures for Cords.	Gold Bronzed.	Leaf Plated.	Iron Border Frames. Bronzed.	Iron Border Frames. Black.
10 inches, round or octagon,	$3.00	$5.25	$6.50	$3.25	$3.50	$11.00	$1.25	$1.00
12 " "	4.00	6.50	7.75	4.25	4.50	15.00	1.55	1.20

SQUARE REGISTERS, WITH PLATED STAR CENTRES.

	No. 1, Best Black.	White Enameled.	Fancy Enameled.	Ventilators, with Fixtures for Cords.	Gold Bronzed.	Leaf Plated.	Iron Border Frames. Bronzed.	Iron Border Frames. Black.
6 by 10 inches, each,	$2.00	$4.00	$5.10	. .	$2.50	$9.00	$1.20	$0.92
8 by 10 " "	2.20	4.62	5.60	. .	2.50	10.00	1.30	1.00
8 by 12 " "	2.70	4.50	6.10	. .	3.10	12.00	1.40	1.16
9 by 14 " "	3.50	4.60	7.25	. .	4.20	15.00	1.60	1.20

ROUND REGISTERS, WITH SLIDE CENTRES.

	No. 1, Best Black.	White Enameled.	Fancy Enameled.	Ventilators, with Fixtures for Cords.	Gold Bronzed.	Leaf Plated.	Iron Border Frames. Bronzed.	Iron Border Frames. Black.
8 inches,	$1.50	$3.30	$4.40	$1.70	$2.00	$8.00	$1.00	$0.75
9 "	1.80	3.30	5.00	2.00	2.25	9.25	1.10	.83
10 "	2.20	4.50	5.50	2.45	2.80	11.00	1.25	1.00
12 "	2.90	5.30	6.50	3.20	3.50	14.00	1.50	1.20
14 "	3.00	6.70	7.50	3.90	4.30	17.50	1.80	1.50
16 "	4.40	7.75	9.25	5.10	5.50	21.00	2.30	1.80
18 "	5.25	8.50	10.00	6.10	6.50	. .	2.85	2.30
20 "	6.75	9.50	11.00	7.15	7.60	. .	3.00	2.50
24 "	9.00	15.00		10.00	11.00	. .	. .	

No.	Page.		
		Ceiling Ventilators, for Large Halls and Churches, 36 inch diameter, bronzed,	$10.00
		" " " " " 60 " "	18.00
		Ornamental Pedestal Registers, with Valves, 16 by 16 inches at base, bronzed, marble tops,	16.50
		" " " " 21½ by 16 " " " "	20.50
		" " " " 21½ by 21½ " " " "	25.50
		Cast Iron Flue Dampers, (Oval,) 4 by 9 inches. Net prices,	.90
		" " " " 4 by 12 " "	1.00
		" " " " 4 by 14 " "	1.13
		Smoke Pipe Registers, 6 inches,	2.25
		" " " 7 "	2.25
		" " " 9 "	2.25
		Ornamental Screens, for Enclosing Coils of Steam Pipes, any length or width. Without Marble Tops—running measure in the clear.	
		Best Gold Bronze, per foot,	3.00
		Best Green " "	2.50
		Marble Tops, 1 inch thick, O. G. Edge, per square foot,	1.10
167	18	Double Acting Force and Lift Pump, 5 inch bore and 14 inch stroke. Finished,	150.00
		" " " " " " " Plain,	125.00
168	"	Hot Water Tanks, with Coil. Furnished to order.	
168½	"	Hot Water Tank. Furnished to order.	
169	"	Cold Water Tanks, with Ball Cocks. Furnished to order.	
170	"	Battlement Bracket, 36 by 10 inches,	3.25
170½	"	Battlement Bracket.	
171	19	Horizontal Tubular Boiler, with Fire-Box. Made and put up to order.	

CLASS FOURTH.—*Continued.*

No. Page.

172 19 VERTICAL TUBULAR BOILERS.

DIAMETER.	LENGTH.	NUMBER OF FLUES.	HEATING SURFACE.	PRICE.
20 inches.	4 feet.	19—2 inches.	28 square feet.	$125.00
24 "	4 "	19—2 "	30 "	150.00
24 "	5 feet 3½ inches.	19—2 "	42 "	162.50
28 "	5 "	37—2 "	72 "	185.00
28 "	6 "	37—2 "	92 "	200.00
32 "	5 "	43—2 "	80 "	250.00
32 "	6 "	43—2 "	103 "	350.00
32 "	10 "	43—2 "	192 "	500.00
36 "	5 "	61—2 "	98 "	350.00
36 "	6 "	61—2 "	120 "	400.00
36 "	8 "	61—2 "	196 "	450.00
36 "	10 "	61—2 "	257 "	550.00
40 "	8 "	66—2 "	198 "	550.00
40 "	10 "	66—2 "	260 "	600.00

173 19 Bath Boilers, Plain or Galvanized. Circulating or Back Log.

LIST OF PRICES.

SIZE.	NUMBER OF GALLONS.	PRICE OF PLAIN.	PRICE OF GALVANIZED.
5 feet by 12 inches.	27 gallons.	$14.50	$22.25
5 " " 10 "	17 "	12.50	18.10
4 " " 12 "	22 "	12.50	19.00
4 " " 10 "	13 "	10.50	15.10
3 " " 12 "	16 "	10.50	15.10
3 " " 10 "	10 "	9.00	12.40
4 " " 14 "	30 "	15.00	22.70
5 " " 14 "	36 "	17.50	27.05
5 " " 16 "	45 "	21.00	32.65
6 " " 14 "	45 "	20.00	30.35
6 " " 12 "	32 "	16.50	25.20
6 " " 16 "	60 "	26.50	40.50
6 " " 18 "	75 "	29.50	44.60
4 " " 16 "	40 "	21.50	34.20
6 " " 21 "	100 "	38.50	60.00
4 " " 6 "	5 "	6.50	9.10

Bath-Boiler Connections extra.

For Back Log Boilers, 1 inch Coupling, consisting of 1 Gooseneck, 1 Ferrule, and 4 foot Tube, at $3.10
For Circulating Boilers, 1 " " " 2 " 2 " " 4 " " " 5.60

174 19 Hot Water Backs. See No. 189, page 26.
175 20 Boiler Fronts, 18 feet by 9 feet 6 inches, without Grate Bars, at $168.50
" " " " " " with 80 Grate Bars, each 4 feet 6 inches, and 2 bearing bars for same, at 327.00
} Any size made to order.

CLASS FOURTH.—*Continued.*

No.	Page		
175½	20	Tubular Boilers, with Steam Drum.	Any capacity made and put up to order.
176	"	Horizontal Tubular Boilers, without Fire Box.	
177	21	Artesian, or other Well Pumps.	

178 " WORTHINGTON STEAM PUMPS. PRICE LIST.

Numbers designating Size.	Diameter of Steam Cylinder in Inches.	Diameter of Plunger in Inches.	Length of Stroke in Inches.	Capacity in Gallons per Minute.	Price.
No. 1.	5¼	3¼	4	10	$ 75.00
" 2.	5	3	6	25	150.00
" 3.	6¾	4½	6	75	225.00
" 4.	12	6	9	125	325.00
" 5.	12	7	9	150	325.00
Second Pattern " "	16	7	9	150	350.00
" 6.	12	10¼	9	400	400.00
do do " "	16	10¼	9	400	475.00
Third do " "	18¾	10¼	9	400	450.00
" 7.	12	14	9	600	525.00
Second do " "	16	14	9	600	550.00
Third do " "	18¾	14	9	600	625.00

N. B.—In case of fire or other emergency, the delivery above stated may be increased to double the quantity.

179 21 WOODWARD STEAM PUMPS.

This Pump is used for supplying Steam-Boilers, Mills, and Public Buildings with water.

In case of fire, it is arranged to discharge any quantity of water, according to size, by simply opening a valve connected to the discharge-outlet. It can also be used for driving a fan or light machinery.

PRICE LIST.

Number.	Diameter of Steam Cylinder in Inches.	Diameter of Water Cylinder in Inches.	Gallons Discharged per minute.	Price.
1	4	2	8 to 12	$100.00
2	5	2½	25 " 39	135.00
3	7	3½	50 " 78	250.00
4	9	5	85 " 129	350.00
5	12	7	167 " 249	400.00
6	12	9	276 " 378	475.00
7	16	9	411 " 513	550.00
8	18	12	730 " 900	650.00
9	20	14	900 " 1200	900.00
10	22	16	1200 " 1500	1100.00
11	24	18	1700 " 2000	1300.00
12	26	20	2000 " 2500	1500.00

CLASS FOURTH.—*Continued.*

GIFFARD'S PATENT SELF-ACTING WATER INJECTORS, (*For feeding Boilers*,) made by William Sellers & Co., Philadelphia, sole Manufacturers and Licensees. For sale by Morris, Tasker & Co.

TABLE OF CAPACITIES.

Size of Injector in Millimetres.	Price at our Works.	Pressure of Steam in Pounds: 10	20	30	40	50	60	70	80	90	100	110	120	130	140	150
		Cubic Feet of Water Discharged per Hour.														
No. 2.	$ [illegible]	[illegible]	6.388	[illegible]	[illegible]	[illegible]	[illegible]	10.511	[illegible]	[illegible]	[illegible]	[illegible]	[illegible]	[illegible]	[illegible]	[illegible]
" 3.	[illegible]	[illegible]	[illegible]	[illegible]	[illegible]	[illegible]	[illegible]	[illegible]	[illegible]	[illegible]	[illegible]	[illegible]	[illegible]	[illegible]	[illegible]	[illegible]
" 4.	[illegible]	[illegible]	[illegible]	[illegible]	[illegible]	[illegible]	[illegible]	[illegible]	[illegible]	[illegible]	[illegible]	[illegible]	[illegible]	[illegible]	[illegible]	[illegible]
" 5.	100.00	[illegible]	[illegible]	[illegible]	[illegible]	[illegible]	[illegible]	[illegible]	[illegible]	[illegible]	[illegible]	[illegible]	[illegible]	[illegible]	[illegible]	[illegible]
" 6.	[illegible]	[illegible]	[illegible]	[illegible]	[illegible]	[illegible]	[illegible]	[illegible]	[illegible]	[illegible]	[illegible]	[illegible]	[illegible]	[illegible]	[illegible]	[illegible]
" 7.	[illegible]	[illegible]	[illegible]	[illegible]	[illegible]	[illegible]	[illegible]	[illegible]	[illegible]	[illegible]	[illegible]	[illegible]	[illegible]	[illegible]	[illegible]	197.67
" 8.	[illegible]	[illegible]	[illegible]	[illegible]	[illegible]	[illegible]	[illegible]	[illegible]	[illegible]	[illegible]	[illegible]	[illegible]	[illegible]	[illegible]	[illegible]	[illegible]
" 10.	[illegible]	[illegible]	[illegible]	[illegible]	[illegible]	[illegible]	[illegible]	[illegible]	[illegible]	[illegible]	[illegible]	[illegible]	[illegible]	[illegible]	[illegible]	[illegible]
" 12.	[illegible]	[illegible]	[illegible]	[illegible]	[illegible]	[illegible]	[illegible]	[illegible]	[illegible]	[illegible]	[illegible]	[illegible]	[illegible]	[illegible]	[illegible]	[illegible]
" 14.	[illegible]	[illegible]	[illegible]	[illegible]	[illegible]	[illegible]	[illegible]	[illegible]	[illegible]	[illegible]	[illegible]	[illegible]	[illegible]	[illegible]	[illegible]	[illegible]
" 16.	[illegible]	[illegible]	[illegible]	[illegible]	[illegible]	[illegible]	[illegible]	[illegible]	[illegible]	[illegible]	[illegible]	[illegible]	[illegible]	[illegible]	[illegible]	[illegible]
" 18.	[illegible]	[illegible]	[illegible]	[illegible]	[illegible]	[illegible]	[illegible]	[illegible]	[illegible]	[illegible]	[illegible]	[illegible]	[illegible]	[illegible]	[illegible]	[illegible]
" 20.	[illegible]	[illegible]	[illegible]	[illegible]	[illegible]	[illegible]	[illegible]	[illegible]	[illegible]	[illegible]	[illegible]	[illegible]	[illegible]	[illegible]	[illegible]	[illegible]

One nominal horse power per hour, will require one cubic foot of water per hour. When the pounds of coal consumed per hour can be ascertained, divide this by 7.5, and the quotient will be the quantity of water in cubic feet per hour.

Minimum capacity of Injectors about [illegible] per centum of tabular capacity.

INSTRUCTIONS FOR FIXING INJECTORS.

1. All the pipes, whether steam, water supply, overflow or delivery, must be of the same internal diameter as the hole in the corresponding branch of each Injector.

2. There must be a cock between the steam space of the boiler and the Injector, one between the bottom of the Injector and the boiler and one in the water supply pipe, when the water gravitates toward the Injector.

3. The supply of steam and water may be regulated by means of the stop-cocks in the steam and water supply pipes, so as to handle the instrument as little as possible. For this reason these cocks should be placed near the instrument.

4. The steam pipe should be taken from the highest part of the boiler, so as to avoid priming, which prevents the efficient working of the Injector. The delivery pipe to be as short and straight as practicable, to reduce friction.

5. The Injector should not be attached to the steam pipe leading to an Engine, unless the pipe should be unusually large, the variation in pressure will break up the jet.

6. It is very important that the water supply pipe should be short and straight where the water does not gravitate to the Injector, for this reason this pipe should be carried vertically down as far as the water level in the supply cistern, which must not be more than five feet below the water chamber in the Injector. In fixing the instrument, place the water chamber of the instrument as low as possible, as the Injector is started most easily under a head.

7. In attaching Injectors to small boilers generating a great quantity of steam, care must be taken to prevent the oscillations in the steam pressure from being suddenly communicated to the Injector, as this will break up the jet, for this reason the opening to the Injector Pipe on Locomotives should not be close to that leading to the cylinders; starting the Engine will cause the Injector to cease working if this difficulty is not guarded against. A good preventive in such cases is to admit the steam to the Injector pipe through numerous small openings in place of one large one.

METHOD OF WORKING.

1. Turn the wheel so as to permit a small quantity of water to flow in the instrument.

2. Open the steam cock connecting the apparatus with the boiler.

3. Turn slightly the handle which will admit a small quantity of steam to the apparatus. A partial vacuum is thus produced, causing the water to enter through the supply pipe.

4. As soon as this happens, which can be observed at the overflow pipe, the supply of steam or water may be increased as required, up to the capacity of the instrument, regulating either by means of the wheel and handle, so as to prevent any overflow.

5. The quantity of water delivered into the boiler, may be varied by means of the stop cocks on the steam and water pipes, without altering the handles on the Injector, a graduated cock on the water supply pipe is very convenient for this purpose.

N. B.—The jamming down of the handles should be avoided; the instrument is purposely made neither steam nor water tight, and when stopped, the steam and water must be shut off with the cocks in the steam and water supply pipes.

No.	Page		
185	[illegible]	Ventilator Fronts, 23¼ by 10 inches,	$1.00
"	"	" " [illegible] by [illegible] "	2.50
"	"	" " 24 by 14 "	2.50
186	[illegible]	Backs and Jambs for Fire Places,	
		Jambs, 24 by 15 inches,	2.50
		" 32 by 12 "	2.25
		" 34 by 12 "	2.75
		" 33½ by 20½ "	4.00
		" 33½ by 28 "	5.25
		Backs, 30 by 24 "	2.00
		" 35 by 27½ "	2.62
		" 35 by 33½ "	3.18

CLASS FOURTH.—*Continued.*

STEAM FITTINGS—HOTEL ARRANGEMENTS, &c.

No.	Page.		
187	26	Cooking Range, for Hotels, &c. Furnished to order.	
188	"	Broiling Oven, 4 feet 4 inches by 4 feet,	$75.00
189	"	Hot Water Backs, 14 inches long, 3½ inches deep, by 8 inches wide,	5.00
		" " 16½ " 4½ " 12 "	7.25
		" " 19 " 4½ " 14 "	9.25
		" " 24 " 4½ " 16 "	12.50
190	"	Hot Water Cylinder, 20 inches outside, 15 inches inside, 10½ inches deep,	24.00
191	"	Hot Water Box, 29 by 29 inches outside, 20 by 19½ inches inside, 11 inches deep,	25.00
192	27	Rotary Roaster, large size,	150.00
		" " small size,	100.00
193	"	Rinsing Tub. Any size made to order.	
194	"	Boiling Tub. " " "	
195	"	Round Tin Steamers, 16 by 15 inches,	13.50
		Oval Tin Steamers, 17 by 23 inches,	14.50
196	"	Wash Tubs. Any size made to order.	
197	28	Force and Lift Pumps,	12.00
198	"	Steam Traps, 7 inches diameter, 1 inch or ¾ inch,	12.00
199	"	Lard Kettle, 3 feet 3 inches by 16½ inches. Sixty gallons capacity,	75.00
200	"	Clothes Wringer. Large size for power,	350.00
		Hand Wringer,	125.00
201	"	Double Steam Kettles, 80 gallons,	90.00
		" " " 65 "	75.00
		" " " 47 "	60.00
202	29	Shaker Washing Machines, with four tubs for power,	325.00
		" " " with six tubs for power,	450.00
		" " " with eight tubs for power,	500.00
203	"	Drying Closets, for Hotels, &c. Constructed to order.	
203a	29a	Mangle,	110.00
203b	29a	Steam Table for Warming Soup, Coffee, Tea and Milk; For Soup, 2 eight gallon Cast-Iron Enameled Kettles, " Coffee, 1 " " Copper, Stone-lined " " Black Tea, 1 seven " " " " " Green Tea, 1 four " " " " " Milk, 1 three " " " "	400.00
204	30	STEAM CARVING TABLE AND DISHES, WITH PLATTERS AND COVERS. For Table 12 feet 10 inches long, with 16 Dishes, Platters and Covers, Slab and Frame, complete,	325.00

STEAM CARVING DISHES, WITH PLATTERS AND COVERS.

NO.	FOR WHAT ADAPTED.	SIZE. Diameter.	PRICE Dish.	PRICE Platter.	PRICE Cover.	TOTAL.
1.	Roast Beef, Roast and Boiled Turkey, &c.	22½ by 15½ inches	$11.00	$7.50	$6.50	$25.00
2.	Corned Beef, Roast Goose, &c.	19½ by 14½ "	8.50	6.00	5.50	20.00
3.	Mutton, Chickens, &c.	17½ by 12½ "	6.00	5.50	4.50	16.00
4.	Various Small Dinner and Breakfast Dishes; Tea, Coffee, Hot Water, &c.	15 by 9 "	4.00	4.50	3.50	12.00

For Table 17 feet 10 inches long, the following sizes are used,

2 of No. 1, at $25.00. 4 of No. 2, at $20.00. 2 of No. 3, at $16.00. 8 of No. 4, at $12.00.

For Table 12 feet long, the following sizes are used. Price, complete, with slab and frame, . . . $242.00

2 of No. 1, at $25.00. 3 of No. 2, at $20.00. 1 of No. 3, at $16.00. 4 of No. 4, at $12.00.

GAS FITTINGS.—PLUMBERS' MATERIALS, &c.

205	31	Gooseneck, 4 inches,	$4.25
		" 3 "	3.50
206	"	Fire Plug, without Case,	18.00
207	"	Fire Plug Case, fitted,	7.00

CLASS FOURTH.—*Continued.*

No. Page.
208 31

GAS DRIPS. TABLE OF PRICES.

Inlet Pipe Connections.	Diameter.	Deep.	Price with Seal.	Price without Seal.
12 inches.	16 inches.	36 inches.	$25.00	$20.00
10 "	16 "	32 "	24.00	20.00
8 "	16 "	27 "	18.00	17.00
6 "	14 "	27 "	15.00	14.00
4 "	10 "	27 "	11.00	10.00
3 "	10 "	22 "	10.00	9.00
2 "	7 "	19½ "	6.50	5.50
1½ "	7 "	19½ "	5.00	4.50

209 " Gas Stops. Long or short pattern.
210 " Water Stops. Long or short pattern.
211 " Gas or Water Street Stop Cases, 4 feet length, 16 by 15 inches at top . . . $15.00

PRICES FOR STOP VALVES FOR GAS AND WATER COMPOSITION SURFACES AND SCREWS.

	Inches.	3	4	6	8	10	12	16	20	24	30
2106 [illegible]	Stop Valves, Inside Screw,	$[illegible]	$[illegible]	$[illegible]	$[illegible]	$50.00	$[illegible]	$166.00	$[illegible]		
2108 [illegible]	Stop Valve, Outside Screw,	[illegible]	[illegible]	[illegible]	[illegible]	60.00	81.00	133.00	200.00	$250.00	$350.00

12 inches and above have Geared Nuts, as shown on figure [illegible]. All stop valves for gas, at [illegible] per cent. discount from above prices.

2106 [illegible] Hydrant Valve.
2108 [illegible] Stop Valve, [illegible] Flanged Ends, [illegible] full size. Composition Surfaces and Screws, Valves fitted up [illegible], Flanged ends Geared, above 12 inches, made to order.

212 32

CAST IRON GAS OR WATER MAIN.

All proved under Hydraulic pressure, 300 lbs. per square inch.

Inside Diameter.	Length.	Weight per Length.	Extra Lengths.	Weight Extra Lengths.	Price per Foot.
12 inches.	9 feet.	610 lbs.	12 ft. 3.21 in.	862 lbs.	Depends on the Iron Market.
10 "	9 "	487 "	12 " 3.08 "	652 "	
8 "	9 "	352 "	12 " 3.51 "	472 "	
6 "	9 "	241 "	12 " 3.56 "	322 "	
4 "	9 "	145 "			
3 "	9 "	102 "			
2½ "	6 "	65 "			
2 "	6 "	40 "			
1½ "	6 "	30 "			

213 " Double Branches,
214 " Single Branches,
215 " Bends,
216 " Steel Hubs,
217 " Reducing Pipes,
218 " Lateral Branches,
219 " Sleeves,
220 " Double Hubs,
221 " Angle Bends,
222 " Caps or Plugs

(213–222:) Any pattern made to order. Standard sizes always on hand. Price depends on the Iron Market.

223 33 Showing Plan of Water Closet Arrangements. Any size made to order.
223½ " Plan of Water Closet Arrangements.
224 } 225 } 34 Single Soil Branches, 8 inches . . . $12.00
" " 6 " . . . 10.50
" " 4 " . . . 9.50
226 " Double Soil " 8 " . . . 15.00
" " 6 " . . . 14.00
" " 4 " . . . 13.50
227 " Reservoir Valve, with strainer, 4 inches . . . 5.00

CLASS FOURTH.—*Continued.*

No.	Page		
228	34	Bath Tubs, with Feet and Waste and Overflow. No. 1, 22 inches wide, 22 inches deep, 5 feet 6 inches long.	$18.00
		" " " " No. 2, 22 by 15 inches wide, 22 inches deep, 5 feet 4 inches long.	18.50
		" " " " No. 3, 22 by 15 inches wide, 19 inches deep, 5 feet long.	12.50
		" " without Waste and Overflow. No. 1, 22 inches wide, 22 inches deep, 5 feet 6 inches long.	16.00
		" " " " No. 2, 22 by 15 inches wide, 22 inches deep, 5 feet 4 inches long.	13.00
	"	" with Feet, without Waste and Overflow. No. 3, 22 by 15 inches wide, 13 inches deep, 5 feet long.	10.50
		" Supply and Waste. New pattern for Asylums, &c.	11.50
229	"	Wash Basins, No. 1, with Feet and Waste.	7.00
		" " " " 1 Cock and 1 Pedestal.	4.74
		" " " " 2 Cocks and 2 Pedestals.	5.55
		" " " " Painted.	3.25
		" " " " " 1 Cock and 1 Pedestal.	4.50
		" " " " " 2 Cocks and 2 Pedestals.	5.80
		" " " " Enamelled.	3.75
		" " " " " 1 Cock and 1 Pedestal.	4.25
		" " " " " 2 Cocks and 2 Pedestals.	5.20
		" " " " " and Painted.	5.00
		" " " " " " 1 Cock and 1 Pedestal.	7.00
		" " " " " " 2 Cocks and 2 Pedestals.	8.75
230	"	Slop Hopper, with Lid. Enamelled and Painted.	11.00
231	"	Soil Pans. Painted, $2.62. Enamelled, $4.00. Plain, $2.50. Wood or Iron Seat, $1.	
232	35	" large Pattern. " 1.38. " 3.00. " 1.25.	
233	"	" with Flanges. " 2.38. " 3.00. " 1.25.	
234	"	" plain Pattern. " 1.38. " 2.00. " 1.25.	
235	"	Urinal Pans. " 1.38. " 2.00. " 1.25.	
236	"	Urinal, with Branch.	7.25
237	"	Half Circle Urinals.	2.50
238	"	Two Basin Sinks, 3 feet 11 inches by 21 inches, by 4½ inches deep. Plain Castings.	7.25
239	"	Corner Sinks, 2 feet 7 inches by 9 inches, by 5 inches deep. Plain Castings.	1.52
		" 28 inches by 9 inches deep. Plain Castings.	1.25
		" 23 inches by 8½ inches deep. Plain Castings.	1.25
240	"	Wash Bowl, 12 inches diameter. Enamelled or Galvanized.	1.25
241	"	Round Cornered Sinks, 3 feet 6 inches by 16 inches, by 4½ inches deep. Plain Castings.	4.25
		" " " 4 feet 6½ inches by 16 inches, by 4½ inches deep. "	5.00
		" " " 24½ inches by 16 inches, by 4½ inches deep. "	1.90
242	"	Sink, with Overflow, 23 inches by 20 inches, by 5½ inches deep. "	2.00
		Round Cornered Sinks, 3 feet 2½ inches by 21 inches, by 5½ inches deep. "	5.00
		" " " 42½ inches by 21 inches, by 5½ inches deep. "	3.25
		" " " 30 inches by 22 inches, by 7 inches deep. "	2.50
		" " " 32½ inches by 21½ inches, by 5½ inches deep. "	2.25
		Galvanized, extra, to be added to weight of Sinks, at per lb.	.03
		Round Cornered Sinks, 27 inches by 17½ inches, by 5½ inches deep. Plain Castings.	2.00
		" " " 24 inches by 17½ inches, by 5½ inches deep. "	1.60
		" " " 23 inches by 18 inches, by 4 inches deep. "	1.12
		" " " 20½ inches by 16 inches, by 4½ inches deep. "	.93
		" " " 24½ inches by 17½ inches, by 4 inches deep. "	1.50
243	"	Large Square Cornered Sinks, 6 feet by 22 inches, by 6 inches deep. "	8.00
244	36	Box Drain Trap. Connections for 6 or 8 inches, 24½ inches by 16½ inches, by 11½ inches deep.	7.00
		" " Connections for 3 or 4 or 5 inches, 20½ inches by 13½ inches, by 10½ inches deep.	4.50
245	"	Belly Traps, 8 inches, plain.	9.00
		" 6 "	5.65
		" 4 "	4.00
246	"	" 1½ inches, with Lid and Bridle to clean, lined.	1.15
		" 1 " " " "	1.75
		" 4 " " " "	3.25
		" 6 " " " "	8.25
		" 8 " " " "	13.50
247	"	Lateral Branch Box Trap. Connections for 6 or 8 inches, 24 by 14½ by 9 inches deep.	12.60
248	"	Right Angle Box Trap. Connections for 8 inches, 20 by 20 by 14½ inches deep.	13.80
249	"	Soil Pan Trap. Connections for 4 inches, 12½ by 9, by 5 inches deep.	3.50
		Right Angle Box Trap. Connections for 4 inches, 14 by 10, by 7 inches deep.	6.50
250	"	S. Trap. Connections for 4 inches.	2.00

CLASS FOURTH.—*Continued.*

No.	Page.		
251	36	U Trap. Connexions for 4 inches,	$2.00
252	"	Soil Pan Trap with Flanges, 12 by 10½, by 4½ inches deep,	2.50
253	"	Urinal Branch. Connexions for 3 inches,	3.50
254	37	Square Gutter Return Bend and Cover, 42½ by 6 inches,	1.20
255	"	Hexagonal Pattern Gutter and Cover, in 6 or 8 feet lengths, at per foot,	.34
		Covers alone, in 6 or 8 feet lengths, at per foot,	.20
256	"	London Pattern Gutter and Cover, in 6 feet lengths, at per foot,	.34
		Covers alone, in 6 feet lengths, at per foot,	.20
257	38	Cast Iron Pipe Screw Joints, 1¼ inches, in 6 feet lengths, at per foot,	.16
		" " " 1½ " 6 " "	.20
		" " " 2 " 6 " "	.22
		" " " 3 " 6 " "	.26
258 } 259 }	[illegible]	Spout Cases, 5 inches by 6 feet,	2.69
		" 4½ inches by 6 feet 6 inches,	4.50
		" 3½ inches by 6 feet,	3.25
		" 3 inches by 6 feet,	3.00
		" 3 inches by 5 feet 6 inches,	2.00

260 "

CAST IRON SMOKE OR SOIL PIPE.

CALIBRE.	LENGTHS.	PER FOOT.
12 inches.	6 feet.	$1.10
10 "	6 "	.85
8 "	6 "	.56
7 "	6 "	.50
6 "	6 "	.40
5 "	6 "	.30
4 "	5 "	.24
3 "	5 "	.20

No.	Page.		
261	"	Lateral Branch for Smoke Pipe. 12 inches, $4.80; 8 inches, $4.50; 7 inches, $3.00; 6 inches, $2.10; 5 inches, $1.80; 4 inches, $1.40.	
262	"	Bend for Smoke Pipe. 12 inches, $6.60; 8 inches, $2.10; 7 inches, $1.25; 6 inches, $1.20; 5 inches, 75 cents; 4 inches, 44 cents.	
263	"	Double Branch for Smoke Pipe. Made to order.	
264	"	Bell Hub for Smoke Pipe. 5 inches, $1.24; 4 inches, 44 cents.	
265	"	Single Branch for Smoke Pipe. 5 inches, 58 cents; 4 inches, 52 cents.	
266	[illegible]	Large Sink, 3 feet long, 2 feet 6 inches wide, 9 inches deep,	12.50
267	"	Horse Trough, Corner Pattern,	2.00
268	"	" Square Pattern,	2.00
269	"	Drip Pan. Of Cast Iron, 3 feet long, with Grates complete, fitted,	25.00
270	"	Half Round Gutter and Cover, 6 feet lengths, per foot,	.32
		Cover alone, 6 feet lengths, per foot,	.18
271	"	Half Round Gutter Bend,	.50
272	"	Half Round Gutter Branch,	.75
273	"	Square Gutter and Cover, 6 feet lengths, per foot,	.50
		Cover alone, per foot,	.18
274	"	Square Gutter Branch,	1.00
275	"	Square Gutter Bend,	.75

PLUMBERS' MATERIALS, &c.—GREEN-HOUSE ARRANGEMENTS.

No.	Page.		
276	39	Tasker's Patent Non-Freezing Hydrant,	10.00
277	"	Stop Cock Box, (Ornamental Lid,) 8 by 8 inches,	1.00
278	"	" " (Plain Lid,) 6 by 6 inches,	.31
279	"	" " (Gas,) 5 by 5 inches,	.24
280	"	Drain Grate, 11¼ by 11¾ inches,	1.50
281	"	Pavement Pipe,	1.50
282	"	Stable Trap. 6 inch Pipe Connexions, 8 by 20½ by 10 inches. Grate, 8½ by 8½ inches,	7.25
283	"	" " 4 inch " " 9¾ by 9½ by 3¾ " " 3¾ by 3¾ "	3.50
284	"	" " [illegible] inch " " 12½ by 5 by 5 " " 4½ by 4½ "	3.25

CLASS FOURTH.—*Continued.*

No.	Page.		
285	40	Plan of Green-House, with Boiler and Pipes complete. These arrangements furnished to order.	
286	41	Wide Pattern Boiler Pipe Stand, 15 inches high, 1 inch roller, 17 inches at top, 2¾ inches at bottom,	$2.87
287	"	Narrow " " 15 " 1 " 11½ " " 6½ " "	1.75
288	"	Expansion Pipe Stand, 2½ inch roller, 2 feet, 8½ inches long,	3.50
	"	" " 2½ " 1 foot 7½ "	3.00
289	"	G. H. Throttle Valve, 4 inches,	5.00
		" " 3 "	4.00
290	"	G. H. Screw Valve, 4 inches,	6.50
291	"	G. H. Double Branch Rib Boiler, 20 inches, fitted. Length of 4 inch pipe for which it is adapted, 500 feet,	30.00
292	"	G. H. Single Branch Rib Boiler, 16 inches, fitted, " 4 " " " 320 "	18.00
293	"	G. H. Double Branch Corrugated Boiler, 28 inches, fitted, " 4 " " " 980 "	63.00
294	"	D. Pattern G. H. Boiler. Large size, " 4 " " " 750 "	100.00
295	42	Smoke Conductor, 3 by 12 inches, in 6 feet 4 inch lengths, at per foot,	1.00
296	"	" " 12¾ by 7 inches, in 9 feet lengths, at per foot,	1.55
297	"	Chimney Top, 6 feet high, 14 inches at top, 21½ inches at bottom,	15.00
298	"	" 16½ inches at top, 16½ inches at bottom,	5.50
299	"	" 30¾ inches high, 9½ inches at top, 11½ inches at bottom,	6.50
300	"	" 18 inches at top, 18 inches at bottom,	5.50
301	43	" 4 feet 2½ inches square, with Corner Pieces,	50.00
302	"	G. H. Double Hub with Outlet, 4 inches,	1.30
303	"	G. H. Pipe with Evaporator, 4 inches, in 9 feet lengths, per foot,	.40
		" " 3 " " " "	.29
304	"	" 4 inches, in 9 feet lengths, per foot,	.36
		" 3 " " "	.25
305	"	G. H. Bend, 4 inches,	1.00
		" 3 inches,	.60
306	"	G. H. Devil Hub, 4 inches,	1.30
307	"	G. H. Single Branch, 4 inches,	1.50
		" " 3 "	1.00
308	"	G. H. Double Branch, 4 inches,	1.85
309	"	G. H. Double Hub with Outlet, 4 inches,	1.70
310	"	G. H. Double Hub Angle Bend, 4 inches,	1.50
311	"	G. H. Return Bend, close pattern, 4 inches,	1.20
		" " " 3 "	.75
		" " medium pattern, 4 inches, (3 inches apart,)	1.40
		" " wide pattern, 4 inches, (6 inches apart.)	1.75
312	44	G. H. Return Hub, 4 inches,	3.25
313	"	Lateral Branch, 4 inches,	1.50
314	"	Angle Bend, G. H.,	
315	"	G. H. H. Branch, 4 inches,	3.00
316	"	G. H. Return Hub, 4 inches,	3.25
317	"	G. H. Three Branch, 4 inches,	3.00
318	"	G. H. Two Branch, 4 inches,	2.35
319	"	G. H. U. Branch, 4 inches,	1.25
320	"	G. H. Reducing Pipe, 4 by 3 inches,	1.00
321	"	G. H. Flange Hub, 4 inches,	1.62
322	"	G. H. Sleeve, 4 inches,	.65
		" 3 "	.49
323	"	G. H. Plug, 4 inches,	.40
324	"	G. H. Cap, 4 inches,	.50
325	"	High Pattern Roller Pipe Stand, 25 inches high, 1 inch rollers, 17 inch top, 9½ inches at bottom,	3.50
326	45	Sylvester Door and Frame, or Fire Front, with Sliding Doors. Frame, 3 feet by 29 inches. Openings, 10¾ by 12, 10¾ by 8¾ inches. Fitted,	15.00
327	"	Man Hole Doors. Frame, 20¾ by 58½ inches. Opening, 4 feet 6 inches by 16½ inches. Fitted,	6.00
328	"	" " " 29 by 19½ " " 22 by 14¼ inches. Fitted,	3.50
329	"	Flue Door, " 8 by 8 " " 5½ by 5½ " "	.75
330	46	Bake Oven Door, " 28½ by 14 " " 21½ by 12 " "	6.00
331	"	Flue Door, " 16½ by 16½ " " 11½ by 11½ " "	2.25
332	"	Bake Oven Door, " 31½ by 23¾ " " 23½ by 17 " "	7.50
333	46	Flue Door, " 15¾ by 13¾ " " 9¾ by 9¾ " "	1.50
334	"	Bake Oven Door, " 24½ by 14 " " 18¾ by 12 " "	4.50

CLASS FOURTH.—*Continued.*

No.	Page	Description	Price
335	46	Fire Door Double Frame, 25 by 12½ inches. Opening, 15 by 12 inches, 13 by 10 inches. Fitted,	$4.00
		" " " 17 by 29 " " 11½ by 11½ inches, 11½ by 9¾ inches. Fitted,	5.00
		" " Small size,	1.50
336	47	Flue Doors. Frame, 19 by 14½ inches. Openings, 15½ by 12 inches. Fitted,	2.00
337	"	" " 10½ by 7 " " 6½ by 5 " "	.75
338	"	" " 15½ by 13 " " 12 by 10½ " "	1.00
339	"	" " 12¾ by 10½ " " 10½ by 8 " "	.87
340	"	Oven Damper " 11½ by 11½ " " 8½ by 8¼ " "	1.00
341	"	Damper, " 18 by 12 " " 7½ by 7½ " "	2.35
342	"	" " 24½ by 10 " " 19 by 4 " "	2.25
343	"	Iron Window Sash. Any size or style made to order.	
344	"	" " " " "	
345	48	" " " " "	
346	"	" " " " "	
347	"	" " " " "	
348	"	Vault Ring and Covers, 18 inches diameter,	4.22
		" " 21 "	4.22
		" " 17 "	1.50
		" " 12 "	1.00
349	"	Vault Covers, 21 inches diameter,	2.00
		" 15 "	1.50
350	"	Vault Grates, 12 "	.84
		" 17 "	1.12
		" 18 "	2.00
		" 21 "	2.88
351	"	Vault Ring and Cover, 30 by 21 inches diameter,	6.50
352	"	Sewer Inlet, 66 by 28½ inches diameter,	20.00
353	"	Box Scrapers,	1.25
354	"	Scrapers,	.37
355	"	"	.25
356	"	Farm Bells. Patterns from 1 to 50 pounds.	
357	"	Door Rollers,	.75
358	"	Garden Rollers, 24 inches diameter, with Balance Handle,	24.00
		" 24 " with Plain Handle,	21.00
		" 24 " Loaded,	23.00
		" 15 " Plain,	12.00
		" 15 " Loaded,	13.00
		" 12 " Plain,	9.00
		" 12 " Loaded,	10.00
358	49	" 30 " With Balance Handle,	38.00
	"	" 30 " " Shafts for Horse,	40.00

359 "

CAST IRON PIPE WITH FACED FLANGES.

Finished with Bolts and Nuts, suitable for Steam Work; proved to 300 pounds pressure per square inch, and intended for 75 pounds working pressure per square inch.

Internal Diameter.	Thickness of Iron of body.	Diameter of Flanges outside.	Diameter of Centre of Holes for Bolts.	Number of Bolts.	Diameter of Bolts.	Length of each piece of Straight Pipe.	Weight of each Middle length.	Price per foot of Length.	Price which is to be added to each length for the pair of ends with bolts.
3 inches.	0.328 in.	6½ in.	5.24	4	⅝ in.	6.0 feet.	93 pounds	$0.28	$1.12
3½ "	0.341 "	7½ "	5.77 5.90	4 or 5	⅝ or ¾ "	8.0 "	112 "	0.33	1.31
4 "	0.354 "	8 "	6.43	5	¾ "	8.0 "	121 "	0.39	1.50
5 "	0.380 "	9 "	7.15 7.62	5 or 6	¾ or ⅝ "	9.0 "	173 "	0.52	1.75
6 "	0.406 "	10½ "	8.82	6	¾ "	11.0 "	237 "	0.66	2.00
8 "	0.458 "	12½ "	10.8	8	¾ "	11.0 "	332 "	1.00	2.75
10 "	0.510 "	15 "	13.16	10	¾ "	11.5 "	511 "	1.37	4.00
12 "	0.536½ "	17½ "	15.50 15.50	10 or 12	¾ or ⅞ "	11.0 "	816 "	1.76	5.25
16 "	0.667 "	24 "	18.9	14	⅞ "	11.0 "	1214 "	2.84	10.50

No.	Page	Description	
360	"	Cast Iron Flange Bends,	At 3 cents per pound, plus the list price for the Flanged Ends with Bolts.
361	"	" " " Single Branches,	(same)
362	"	" " " Double Branches,	(same)

CLASS FOURTH.—*Continued.*

No.	Page		
363	50	Vault Frame and Covers.	
364	"	Carpet Winding Machine. Size of Wheel, 15¾ inches diameter. Legs, 18¾ inches long, 2½ inches high. Height to top of Wheel, 41 inches. Height to top of Spindle, 12 inches,	9.00
365	"	Umbrella Stands,	3.50
366	"	Brush Scrapers,	2.75
367	"	Pan Scrapers,	1.25
368	50	Garden Chairs, Plain,	5.25
		" Bronzed,	6.00
369	"	" Plain,	8.50
		" Bronzed,	9.00
370	"	Double Garden Chair, Plain,	14.00
		" " Bronzed,	15.00
371	"	Garden Stool, Plain,	1.25
		" Bronzed,	1.75
372	"	Table Legs,	.75
373	"	Umbrella Stands,	2.50
374	"	Mirror Frames,	2.00
375	"	" "	1.00
376	"	Spittoon,	.50
377	"	Boot Jack,	.50
378	51	Stairway Brackets, 24 inches length or projection, 9 inches wide, 3 inches wall bearing,	1.25
379	"	Mantel Brackets, 26½ " " 8 " 3½ by 3½ inches wall bearing,	1.00
380	"	" 17 " " 11 " 3 inches wall bearing,	.75
381	"	" 17 " " 10 " 3 " "	.50
382	"	" 10 " " 9 " 3 " "	.50
383	"	Horse Head Hitching Post,	4.00
384	"	Octagonal Hitching Post,	6.00
385	"	Ornamental Hitching Post,	3.25
386	"	Slender Pattern Hitching Posts,	3.25
387	52	Gas Bracket, 1 feet 9½ inches length or projection, 21 inches wide, 14 by 12 inches wall bearing,	16.00
388	"	Gallery Bracket, 22 inches length or projection, 12 inches wide, 14½ by 11½ by 4 inches wall bearing,	2.00
389	"	" " 16 " " " 9 " " 6 by 6 " " "	1.25
390	"	" " 21 " " " 21 " " 6½ by 5 " " "	2.00
391	"	Window " 13 " " " 6 " " 6 by 5½ " " "	.50
392	53	Capital " 4 feet length or projection, 24 inches wide, 6 inches wall bearing,	2.50
393	"	Gallery " 19 inches length or projection, 10 inches wide, 5 by 3 inches wall bearing,	1.25
394	"	Mantel " 9 " " " 11½ " " 3 inches wall bearing,	.50
395	"	" " 17 " " " 13 " " 3 " " "	1.00
396	"	" " 11½ " " " 11½ " " 2 " " "	.50
397	"	Pedestal Lamps, 4 feet 2 inches high, (with any Pattern Lantern purchaser may desire, extra,)	3.00
398	54	" " and Lantern, 7 feet 6½ inches high from base to bottom of Lantern,	20.00
		" " without " " " " " " "	19.00
399	"	Pedestals for Lantern, 5 feet 9 inches high from base to bottom of Lantern,	6.00
		" with " " " " " " "	10.50
400	"	Park Lamp Posts for Lantern, 6 feet 2 inches high from base to bottom of Lantern,	6.00
		" " " with " " " " " " "	10.50
401	"	Street Lamp Posts for Lantern, 5 feet 10 inches high from base to bottom of Lantern,	6.00
		" " " with " " " " " " "	10.50
402	"	" " " for " 8 feet 7 inches high from base to bottom of Lantern,	8.00
		" " " with " " " " " " "	12.50
403	"	Street Lanterns, 24 inches high,	4.50
403a	54a	Gothic Lanterns, 17 inches by 3 feet,	6.00
403b	54a	" " 17 " 2 "	4.00
403c	54a	" " 17 " 2 " 10 inches,	5.00
404	54	Gallery Lamp Posts, 6 feet 9 inches high,	3.00
		" " " 4 feet 2 inches "	2.00
		" " " 4 feet 3 inches "	2.00
405	55	Awning Posts. Caps 50 cents each, Bases $1,00 each,	
405a	54a	" " " 40 " " 1.50 "	

CLASS FOURTH.—*Continued.*

No. Page

[illegible] [illegible] Awning Posts. Caps [illegible] cents each, Bases 32 cents each.
[illegible] [illegible] " " " 62 " " 55 " "
[illegible] [illegible] " " " 86 " " 55 " "
[illegible] [illegible] " " " [illegible] " " $1,00 " "
[illegible] " Gas Pipe Railing.
[illegible] " Ornamental Gas Pipe Railing.
[illegible] [illegible] Balcony Rail of Gas Pipe. Any size made to order.
[illegible] " Railroad and Turnpike Gates of Gas Pipe. Any size made to order.
[illegible] [illegible] Garden or Park Railings of Gas Pipe.
[illegible] [illegible] Church or Cemetery Railings of Gas Pipe.
[illegible] [illegible] Spiral Stair Case of Cast Iron, consisting of Step Riser and Rail. Price per step. . . . $4.00
Any Pattern can be got up at short notice.
[illegible] [illegible] Spiral Stair Case or Fire Escape. Any number of Steps furnished to order at $[illegible] each step, including Riser and Hand-Rail, &c. Top and Bottom Platforms extra, $[illegible].
[illegible] [illegible] Cast Iron Column, 1½ inches at neck, 4½ inches at base. Made any thickness or length.
[illegible] " " " 1½ " " [illegible] " " " " "
[illegible] " " " 2 " " [illegible] " " " " "
[illegible] " " " 4 " " [illegible] " " " " "
[illegible] " " " [illegible] " " [illegible] " " " " "
We also have Patterns for a large variety of other Columns.
[illegible] [illegible] Showing Circular Hay Rack made of Gas Pipe.
[illegible] " Showing Stall Division. Cast Iron Post Top and Bottom Rail.
[illegible] [illegible] Showing Stable Arrangements of Gas Pipe.
We would call the attention of Builders and others to this economical as well as durable mode of fitting up stables. We are prepared to give estimates for this kind of work to parties furnishing us with plans of Stables, the amount of stall room required for each animal, &c.
[illegible] [illegible] Jackson's Patent Governor Valves. Furnished to order.
[illegible] " Cast Iron Throttle Valves, 3 inches . . . 2.50
[illegible] " Slide Wind Valves, 3 inches, . . . 2.50
[illegible] " Throttle Wind Valves, 3 inches . . . 1.50
[illegible] " Spring Valves for Water Closets. Complete, . . . 8.50
[illegible] " Pulleys for Spring Valves. . . . 1.00
[illegible] [illegible] Tasker's Patent Self-Regulating Hot Water Furnace. Designed for Warming Public and Private Buildings, Hospitals, Churches, Conservatories, School and Green Houses.
Persons desirous of knowing the cost, will please furnish us with plans of the building proposed to be warmed.
[illegible] [illegible] Single Retort House Gas Apparatus, (Capacity, 1000 feet in six hours,) with Brick Lining and all connections to Holder, . . . 200.00
Gas Holder, say 1000 feet capacity, punched and ready to go together; delivered in Philadelphia, . . . 250.00
Frame, Columns, Weights, Chains, Pulleys and Counter Balance Weights, &c., &c., . . . 170.00

Total cost, . . . 620.00

[illegible] [illegible] Double Retort House Gas Apparatus, (capacity, 2000 feet in six hours,) with all connections to Holder, Brick Lining, &c., . . . 300.00
Gas Holder, say 2000 feet capacity, punched and ready to go together; delivered in Philadelphia; with Frame, Columns, Weights, &c., complete, . . . 700.00

Total cost, . . . 1000.00

From our experience of many years, we can recommend our "House Gas Apparatus" as being the very best and most economical in use.

[illegible] [illegible]
[illegible] [illegible] } Coal Gas Works.
[illegible] [illegible]

We are prepared to furnish all the Cast and Wrought Iron work for Coal Gas Works, of any size and capacity, and plans for Retort and Purifier Houses, or to contract with companies to erect the works and furnish everything necessary to deliver the gas to the consumer.

77, 78 Malleable Iron Fittings, per pound. . . . 11

[illegible] GAS COCKS.

Inches,	¼	⅜	½	¾	1	1¼	1½	2
Per Hundred,	$0.40	[illegible]	[illegible]	[illegible]	$0.75	[illegible]	[illegible]	[illegible]

[illegible] [illegible] Combined Drip and Valve for Gas Works, for [illegible] . . . [illegible]

CLASS FOURTH.—*Continued.*

No.	Page.		
486	79	Poker, }	
487	"	Chisel Bar, }	$15.00
488	"	Retort Scraper, }	
489	"	Tongs,	1.25
490	"	Coke Barrow,	55.00
491	"	Coal Barrow for 8 foot Retorts,	100.00
		" " 7 " "	75.00
492	79	Charcoal Barrow, with Lid,	70.00
493	80	Plan of Rosin Gas Works. Retorts set in Brick Work.	
494	81	Gas Pipe Rolling and Steam Radiator combined.	
495	81	Chemical Retort, 3 feet 2½ inches high by 24 inches, 1 inch thick,	25.00
		" 18 inches high by 11¾ inches, ½ inch thick,	6.50
496	82	Ornamental Garden Vase,	9.00
497	"	Garden Vase,	8.00
[illegible]	"	"	6.00
		Square Pedestal for Small Garden Vase,	2.50
511	86	Charcoal Barrow, 5 feet 6 inches by 2 feet 9 inches wide, by 2 feet deep,	80.00

CLASS FIFTH.

GAS AND STEAM FITTERS' TOOLS. *Nett Cash.*

104	10	Screwing Stocks, with Solid Dies. No. 1, for ¼ inch, ⅜ inch, ½ inch Tube,	8.50
		" " " " No. 2, for ¾ inch, 1 inch Tube,	9.50
		" " " " No. 3, for 1¼ inch, 1½ inch, 2 inch Tube, with Driving Screw,	17.00
		" " for Cutting Screws. No. 4, for 2½ inch, 3 inch, " " "	30.00
		" " " " No. 5, for 3½ inch, 4 inch, " " "	45.00
105	"	Cutting-off Stocks. No. 1, for ¼ inch, ⅜ inch, ½ inch, ¾ inch, 1 inch Tube,	5.00
		" " No. 2, for 1¼ inch, 1½ inch, 2 inch Tube,	6.00
		" " No. 3, for 2½ inch, 3 inch Tube,	15.00
		" " No. 4, for 3½ inch, 4 inch Tube,	20.00
106	"	Ratchet Wrench,	10.00
107	"	Vice Clamp,	4.50
108	"	Drill Stock and Clamp, for Tapping Street Mains. From 3 to 1½ inches, $6.00; from 6 inches to 3 inches, $10.00; from 12 inches to 6 inches, $16.00,	
109	"	Stock and Dies, for Brass Pipe,	7.50
110	"	Solid Dies, Left Hand or Right Hand Threads. ⅛ inch, ¼ inch, ⅜ inch, ½ inch. Each,	1.75
		" " " " ¾ inch, 1 inch. Each,	2.50
		" " " " 1¼ inch, 1½ inch, 2 inch. Each,	3.25
		" " " " 2½ inch. Each,	4.25
		" " " " 3 inch. "	5.50
111	"	Gas Fitters' Universal Clamp Vice, Movable Head,	10.00

		Inches,	⅛	¼	⅜	½	¾	1	1¼	1½	2	2½	3	3½
112	"	Taps,	$0.90	$0.90	$1.15	$1.25	$1.50	$1.75	$2.00	$2.50	$3.25	$6.00	$8.00	.
113	"	Reamer,	.90	.90	1.15	1.25	1.50	1.75	2.00	2.50	3.25	6.00	8.00	.
114	"	Drills,	.	.60	.65	.75	.85	.95	1.00	1.10	1.40	.		.
115	"	Outside Chasers,	.	.75	.75	.75	.75	.75	.	.	.	.		.
116	"	Inside Chasers,	.	.75	.75	.75	.75	.75	.	.	.	.		.

CLASS FIFTH.—*Continued.*

No.	Page.		
117	10	Tap Wrench,	$1.50
118	11	Breast Drill,	5.00
119 } 120 }	"	Wall and Cold Chisels, } per pound,	.50
123 } 124 }	"	Cap and Caulking Chisels, } per pound,	.50
121	"	Single Geared Hand-Screwing Machine, with Universal Gripping Chuck, Cutting-off Head and Solid Dies, for ¼ inch, ⅜ inch, ½ inch, ¾ inch, 1 inch, 1¼ inch Tube,	75.00
122	"	Double Geared Hand-Screwing Machine, with Universal Gripping Chuck, Cutting-off Head and Solid Dies, for ½ inch to 2 inches Tube, inclusive,	150.00
		Double Geared Screwing Machine, with Pulleys for Power, with Universal Gripping Chuck, Cutting-off Heads and Dies, for ¼ inch to 2 inches Tube, inclusive,	175.00
125	12	Brass Drip Pump,	18.50
126	"	Gas Fitters' Proving Pump and Gauge, No. 1,	19.00
		" " " " No. 2,	21.00
		" " " " No. 3,	16.00

		Inches,	9	12	16	19	24
127	"	Cylinder Wrenches (American), bright finished,	$1.38	$1.67	$2.38	$2.87	$4.00
		" " black unfinished,	1.25	1.75	2.25	2.75	3.75

		For	⅛	¼	⅜	½	¾	1	1¼	1½	2	2½	3	3½	4	4½	5	6	inch Tube.
128	"	Pipe Tongs,	$1.00	$1.00	$1.00	$1.25	$1.50	$1.75	$2.00	$2.50	$3.00	$4.00	$5.00	$6.00	$7.00	$8.50	$10.00	$14.00	Price per pair.

		Number,	00	0	1	2	3	4	5
129 } 130 } 131 } 132 }	"	Steel Burner and Meter Pliers,	$0.63	$0.75	$1.25	$1.50	$1.75	$2.00	$2.50

133 12 Floor Chisels, $1.50.

134 }
135 } " Plaster and Cold Chisels, 50 cents per pound.

137 " Blow Pipes, 62 cents.

138 " Torches, 62 cents.

		For	½	⅝	¾	⅞	1	1¼	1½	2	inch Tube.
136	"	Augers,	$0.62	$0.75	$0.87	$1.25	$1.50	$1.75	[illegible]	$2.50	Price, each.

515	56	Proving Pump for Street Mains, complete,	$50.00
516	"	Gilmore's Patent Adjustable Tongs. For ¼ to ⅝ inch Pipe, per pair,	3.25
		" ½ to 1¼ " " " "	4.00
517	"	Brown's Patent Pipe Tongs. No. 1, For ⅜ to 2½ inch Pipe, per pair,	5.00
		No. 2, " ⅛ to 1¼ " " " "	3.75
		No. 3, " ⅛ to 1¼ " " " "	2.00
518	"	Brown's Patent Pipe Cutter No. 1, cutting from ⅛ to 1¼ inch pipe, per pair,	3.00
		No. 2 " 2 to 1 " " "	5.00
		Stanwood's Patent Pipe Cutters, each,	5.00

CLASS SIXTH.

ARTESIAN WELL TUBES, TOOLS, PUMPS, &c.

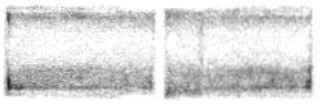

No. Page.
102 9

CAST IRON ARTESIAN WELL PIPE, WITH BOLLES' PATENT FLUSH JOINTS.

Nominal Diameter.	Inside Diameter at Joint.	Outside Diameter at Joint.	Length of each Piece.	Price per Foot, with Band, including Patent Fee.	Extra Price of Bottom Pipe and Band.	Weight per Piece.	Weight of Band.
Inches.	Inches.	Inches.	Feet.	$ c.	$ c.	Pounds.	Pounds.
12	12⅞	13⅞	10	[illegible]	4.50	600	34
10	10	11½	9	1.88	2.87	[illegible]	23
8	8	9¼	8	1.75	2.25	298	14
6	5¾	6¾	8	[illegible]	1.82	[illegible]	10

103a 9

WROUGHT IRON ARTESIAN WELL TUBE, FLUSH JOINTS.

Length of each Piece from 10 to 15 Feet.

Inside Diameter.	Outside Diameter.	Weight per Foot.	Price per Foot, including Joints.	Extra Price for Steel End at Bottom.
Inches.	Inches.	Pounds	$ c.	$ c.
2½	2¾	5.8	0.60	1.25
3	3½	7.7	.85	1.40
3½	4¼	9.7	1.05	1.55
4	4¾	11.8	1.20	1.70
4½	5	12.9	1.60	1.85
5	5 [illegible]	14.5	1.80	2.00
6	6⅝	19.8	2.85	2.30
7	7⅝	24.6	4.50	2.65
8	8⅝	29	5.50	3.00

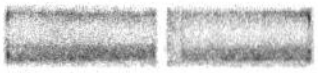

103b 9

LIGHT WROUGHT IRON ARTESIAN WELL TUBE, FLUSH JOINTS.

Length of each Piece from 10 to 15 Feet.

Outside Diameter.	Thickness of Tube.	Weight per Foot.	Price per Foot, including Joints.	Extra Price for Steel End at Bottom.
Inches.	Inches.	Pounds.	$ c.	$ c.
2½	0.109	2.6	0.41	1.25
2¾	0.109	2.98	.46	1.33
3	0.132	[illegible]	.52	1.40
3¼	0.12	3.87	.57	1.48
3½	0.12	4.21	.68	1.55
3¾	0.12	4.39	.78	1.62
4	0.134	[illegible]	.81	1.70
4½	0.134	[illegible]	1.10	1.85
5	0.148	[illegible]	1.27	2.00
6	0.165	10	[illegible]	2.30
7	0.165	12	2.45	2.65
8	0.165	14	2.90	3.00

CLASS SIXTH.—*Continued.*

No. Page.

WROUGHT IRON ARTESIAN WELL PIPES, SOCKET JOINTS.

1035 9

Finished smooth inside and cut especially for this use.

Inside Diameter.	Outside Diameter.	Weight per foot.	Price per foot, including Socket.	Extra price for Steel End at bottom.
Inches.	Inches.	Pounds.	\$ c.	\$ c.
2½	2⅞	5.8	.60	1.25
3	3½	7.7	.85	1.40
3½	4⅛	9.7	1.05	1.55
4	4⅝	11.8	1.30	1.70
4½	5	13.9	1.60	1.85
5	5 9/16	14.5	1.80	2.00
6	6⅝	19.8	2.85	2.30
7	7⅝	24.6	4.50	2.65
8	8⅝	29	5.50	3.00

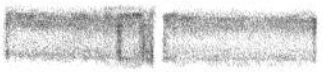

LIGHT WROUGHT IRON ARTESIAN WELL TUBE, SOCKET JOINTS.

1034 8

Finished smooth inside and outside especially for this use.

Outside Diameter.	Thickness of Metal.	Weight per foot.	Price per foot, including Socket.	Extra price for Steel-end at bottom.
Inches.	Inches.	Pounds.	\$ c.	\$ c.
2¼	0.109	2.6	.42	1.25
2½	0.109	2.98	.46	1.32
3	0.109	3.16	.52	1.40
3¼	0.12	3.87	.52	1.48
3½	0.12	4.21	.68	1.55
3¾	0.12	4.90	.72	1.62
4	0.12	5.23	.84	1.70
4½	0.134	5.94	1.10	1.85
5	0.148	6.48	1.32	2.00
6	0.165	10	1.91	2.30
7	0.165	12	2.45	2.65
8	0.165	14	3.09	3.00

FLUSH JOINT ARTESIAN WELL TUBES ARE MADE UNDER THE PATENT OF J. N. BOLLES, JUNE 19th 1855.

Claim.—"The mode of rendering Cylinders or Tubes flush or upon a line on their exterior surfaces, for Artesian Wells, or for other purposes, as described, or any other mode, substantially the same, which will produce the same effect. Patented, June 19, 1855."

The above Patent is owned by Morris, Tasker & Co.; they grant no licenses to others to manufacture, or for the use of any but their own manufacture.

Also to connect to the above tubing, Pumping Cylinders, of all diameters; Composition or Cast Iron Chambers; Composition Buckets (not requiring packing) and Valves, and all the working parts very substantial for hard service in deep well pumping.

ARTESIAN WELL PUMPS, WITH WROUGHT IRON BRASS LINED CHAMBERS, BRASS PLUNGER AND LOWER BUCKET GLOBULAR SEAT POPPET VALVES.

Any size manufactured to order.

Internal Diameter, (Nominal.)	Internal Diameter, (Actual.)	Stroke.	Price.	Increased Length of Stroke, per foot, extra.
Inches.	Inches.	Feet.	\$ c.	\$ c.
2½	2⅝	4	31.00	1.50
3	2⅞	5	41.25	2.00
3½	3⅜	6	56.50	2.75

CLASS SIXTH.—*Continued.*

ARTESIAN WELL BORING TOOLS.

No.	Part			6 ft.	8 ft.	10 ft.	12 ft.	15 ft. Long	
800	84	Boring Rods, 1½ inch Square Iron for Sinking Wells, 4 to 12 inches internal diameter, at $4.00 per end, and 50 cents per foot of length.	1½ in. square.	$7.00	$8.00	$9.00	$10.00	$11.50	Per piece. Complete.
801		Boring Rods, ⅝ inch Square Iron for Sinking Wells; 2 to 2½ inches Internal diameter, at $3.25 per end, and 35 cents per foot of length.	⅝ inch do.	5.35	6.05	6.75	7.75		Per piece. Complete.

No.	Part	INTERNAL DIAMETER OF WELL.	2	2½	3	3½	4	5	6	8	10	12	Inches
			$ c.	$ c.	$ c.	$ c.	$ c.	$ c.	$ c.	$ c.	$ c.	$ c.	
802	"	Pod Auger,	5.25	6.25	7.25	8.25	9.25	11.25	13.25	17.25	21.25	24.25	With Steel Cutting Edges
803	"	Lip "	5.25	6.25	7.25	8.25	9.25	11.25	13.25	17.25	21.25	24.25	
804	"	Worm " (8 twists,)	7.25	8.75	10.25	11.75	13.25	16.25	19.25	25.25	31.25	37.25	
805	"	" " (4 ")	6.00	7.12	8.25	9.38	10.50	12.75	15.00	19.50	24.00	28.50	Solid Steel Worm, below 4 inches; Steel Pointed, above 4 inches.
806	"	" " (2 ")	5.00	6.00	7.00	8.00	9.00	11.00	13.00	17.00	21.00	24.00	
807	"	" " (cone.)	4.75	5.62	6.50	7.38	8.25	10.00	11.75	15.25	18.75	22.25	
808	"	" " (taper.)	4.75	5.62	6.50	7.38	8.25	10.00	11.75	15.25	18.75	22.25	
809	"	Spiral " (flat,)	4.75	5.62	6.50	7.38	8.25	10.50	12.75	17.25	21.75	26.75	
810	"	" " (do.)	4.75	5.62	6.50	7.38	8.25	10.50	12.75	17.25	21.75	26.75	Solid Steel Worm
811	"	Y		2.25					3.25				
812	"	Cross Head,	4.00	4.75	5.50	6.25	7.00	8.50	10.50	13.00	16.00	19.00	
813	"	Wrench Handle,		11.00					15.40				
814	"	" Bar,		2.25					3.00				
815	"	Lifter,		3.95					5.00				
816	"	Hook,		1.75					2.50				
817	"	Well Hook,		4.50					6.00				
818	"	Combination Auger,		not made			58.00	67.50	75.00	83.00	92.00	100.00	Triplicate Steel Cutters.
819	85	Boulder Cracker,	5.50	8.50	11.50	14.50	17.50	23.50	29.50				
820	"	Spring Catch,	4.50	5.25	6.00	6.75	7.50	9.00	10.50	13.50	16.50	19.50	Steel Wires.
821	"	Hook,		2.75									
822	"	"		.50									
823	"	Drill Stock and Drill for soft rock,	17.50	20.00	22.50	25.00	27.50	32.50	37.50	47.50	57.50	67.50	Triplicate Steel Drills.
824	"	Catchall,	47.50	48.75	50.00	51.25	52.50	55.00	57.50	62.50	75.00	85.00	4 pairs Steeled Bits.
825	"	"	40.00	41.25	42.50	43.75	45.00	47.50	50.00	55.00	60.00	75.00	
826	"	Catch Hook,	5.00	5.25	5.50	5.75	6.00	11.00	13.00	17.00	21.00	24.00	
827	"	Hook,		2.25									
828	"	Drift Hammer,											
829	"	Thief Bucket,	6.00	7.12	8.25	9.38	10.50	12.75	15.00	19.50	24.00	28.50	
830	"	Socket and Hook,			7.00				9.00				
831	"	Tubular Drill Rods,	10 cts. per foot, 1½ in. outside diameter.										
832	"	Drill Stocks,	8.00	9.50	11.00	12.50	14.00	17.00	20.00	26.00	32.00	38.00	
833	"	Drills,	8.00	9.50	11.00	12.50	14.00	17.00	20.00	26.00	32.00	38.00	
834	"	"	3.50	4.12	4.75	5.37	6.00	7.25	8.50	11.00	13.50	16.00	
835	"	Clamps,							10.00				
836	86	Catch for hauling Pipe,											
837	"	Gin Pulleys,	12 in. wheel, $12.50; 10 in. wheel, $10.00.										
838	"	Section of a Pump'g Cylinder,											
839	"	Deep Well Pump,											

840 " **Apparatus for Sinking Artesian Wells, consisting of Derricks with Sheers, etc., complete as shown in drawing, exclusive of Ropes and Chains, $375.00**
497 83 **Bolles' Rock Drilling Machine.**

ANY SIZE DERRICKS MADE TO ORDER AT LOWEST RATES.

LAP WELDED AMERICAN CHARCOAL IRON BOILER FLUES.

No 1

With Plain Ends

No 2

With Safe Ends

No 3

With Safe Ends 1/16 inch larger than rest of the Flue

No 4

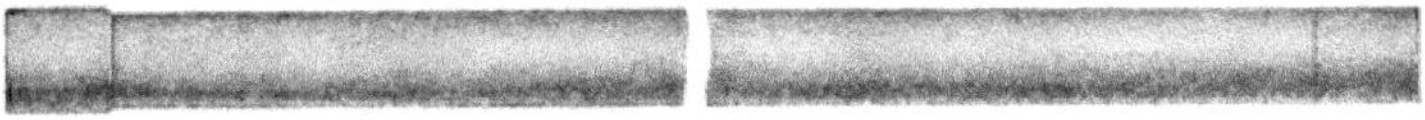

With Safe Ends one of which is 1/16 in larger than rest of the Flue

No 5

With Safe Ends one of which is 1/[illegible] in less than rest of the Flue

No 6

Same with round Shoulder

No 7

With Screw on one End and Collar on the other

No 8.

Cross Flue for Fire Box

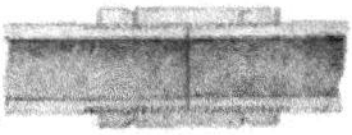

Section

Plate 2

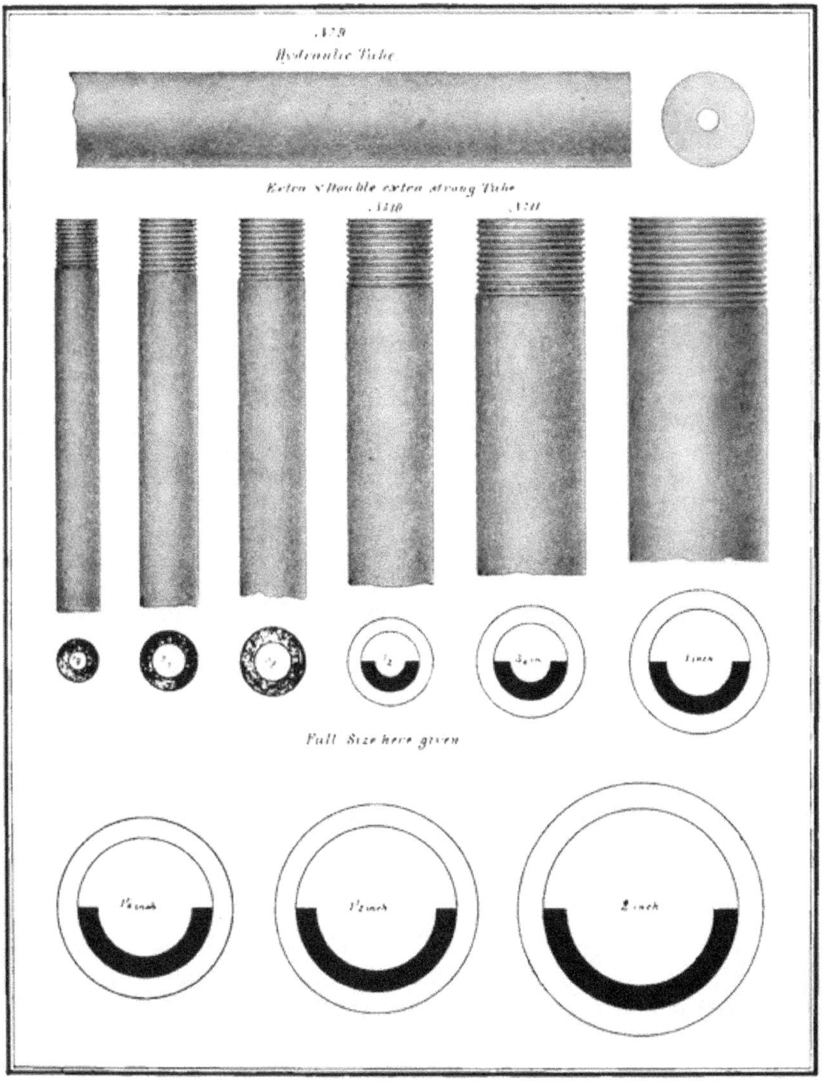
Hydraulic Tube
Extra & Double extra strong Tube
3/4 inch
1 inch
Full Size here given
2 inch

MORRIS, TASKER & Co's ILLUSTRATED CATALOGUE.

Plate 3

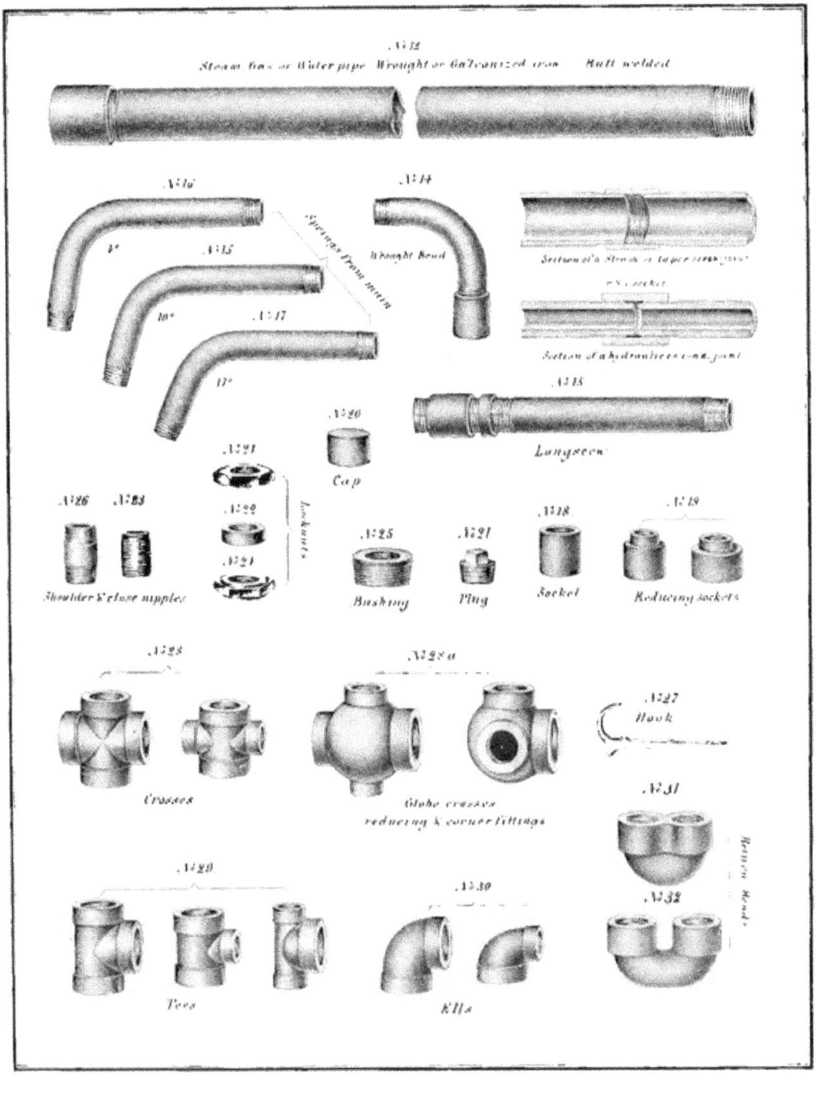

MORRIS, TASKER & C°'s ILLUSTRATED CATALOGUE.

Plate 4

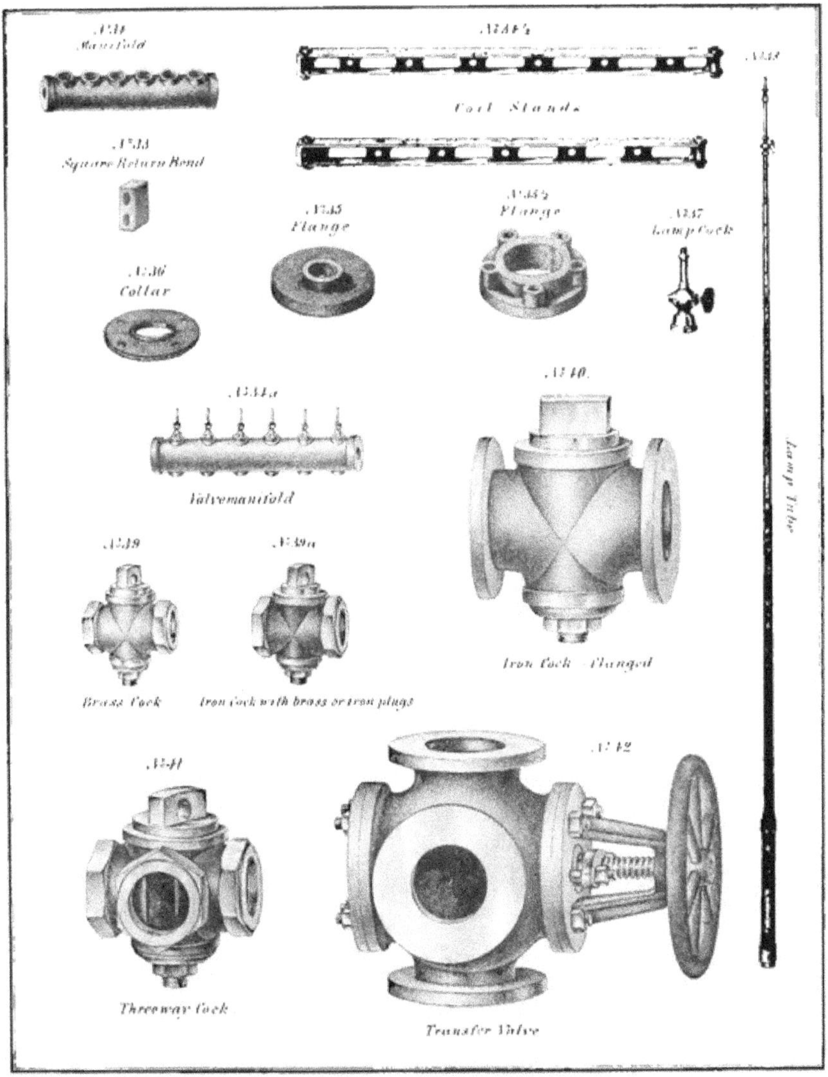

Plate 5

No. 43

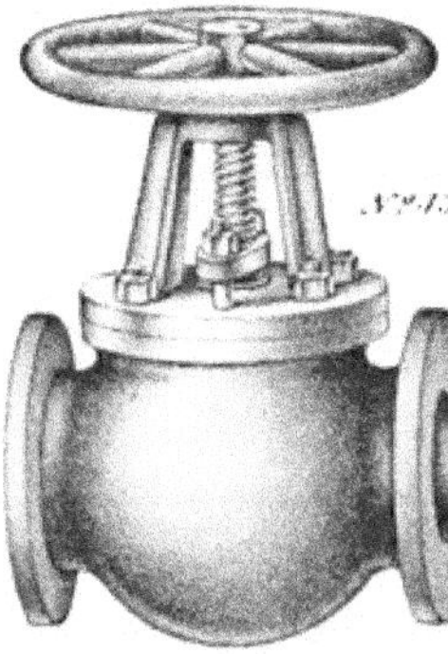

Flanged Globe Valve

No. 43 a

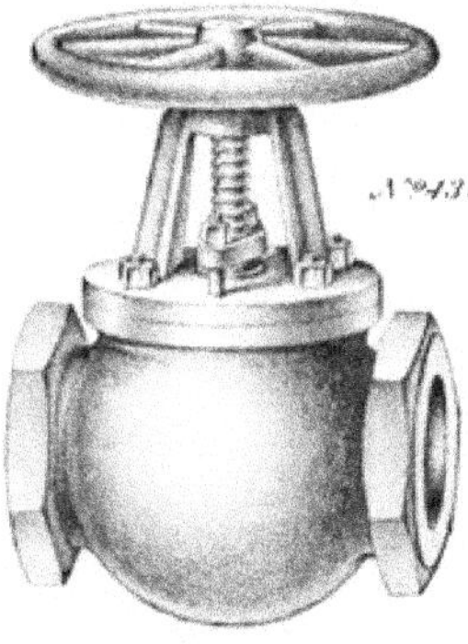

Screwed Globe Valve

No. 43 b

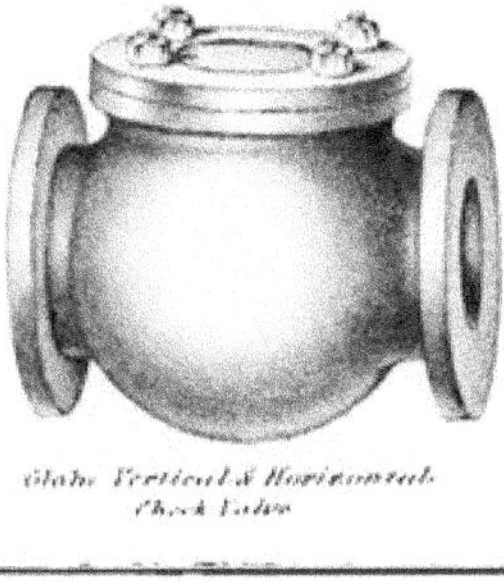

Globe Vertical & Horizontal Check Valve

No. 43 c

Globe Safety Valve

No. 43 d

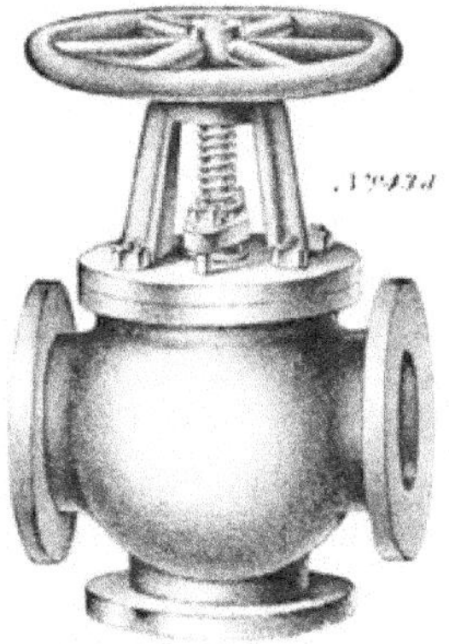

Globe Cross Valve

No. 43 e

Globe Back pressure Valve

MORRIS, TASKER & Co.'s ILLUSTRATED CATALOGUE.

Plate 1a

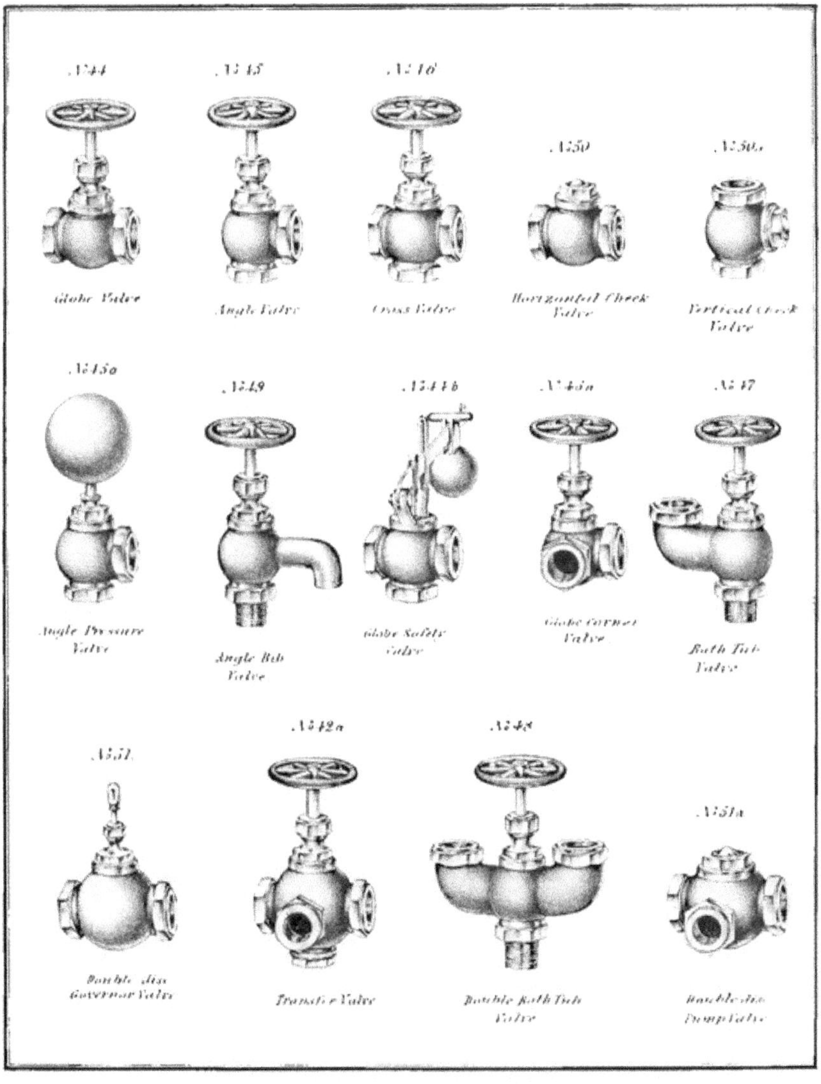

Plate 6

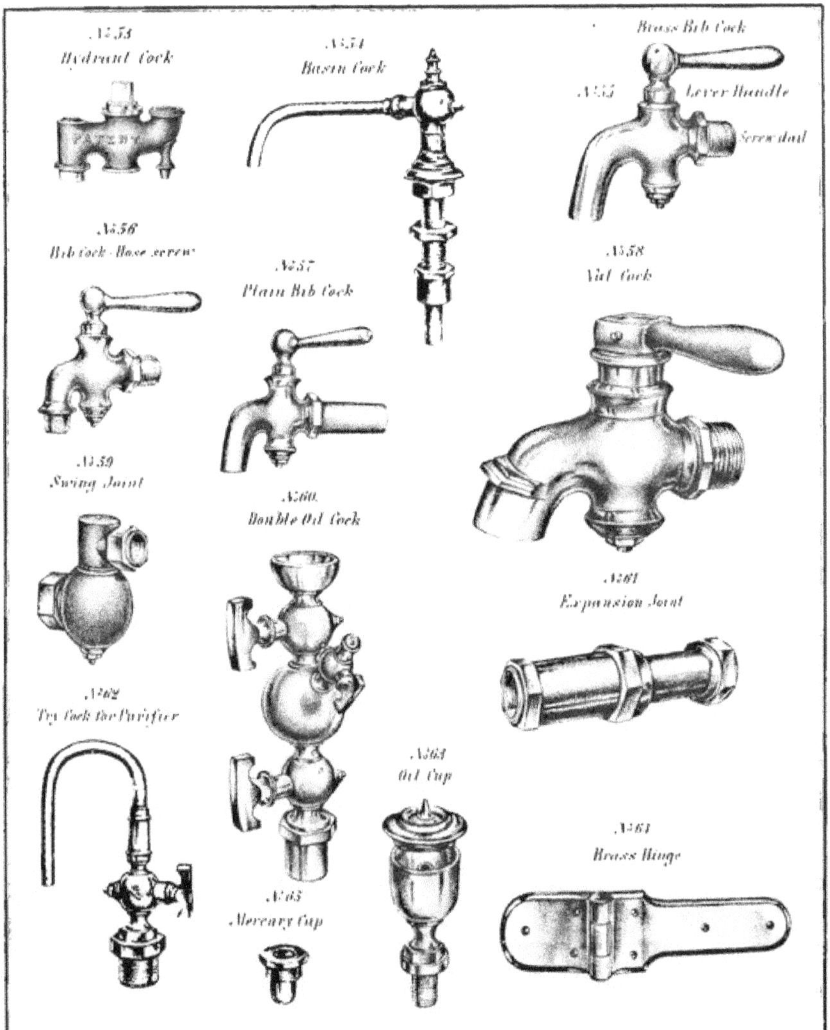
No 53
Hydrant Cock
No 54
Basin Cock
Brass Bib Cock
No 55
Lever Handle
No 56
Bib cock-Hose screw
No 57
Plain Bib Cock
No 58
Vat Cock
No 59
Swing Joint
No 60.
Double Oil Cock
No 61
Expansion Joint
No 62
Try Cock for Purifier
No 63
Oil Cup
Mercury Cup
No 64
Brass Hinge

Plate 7

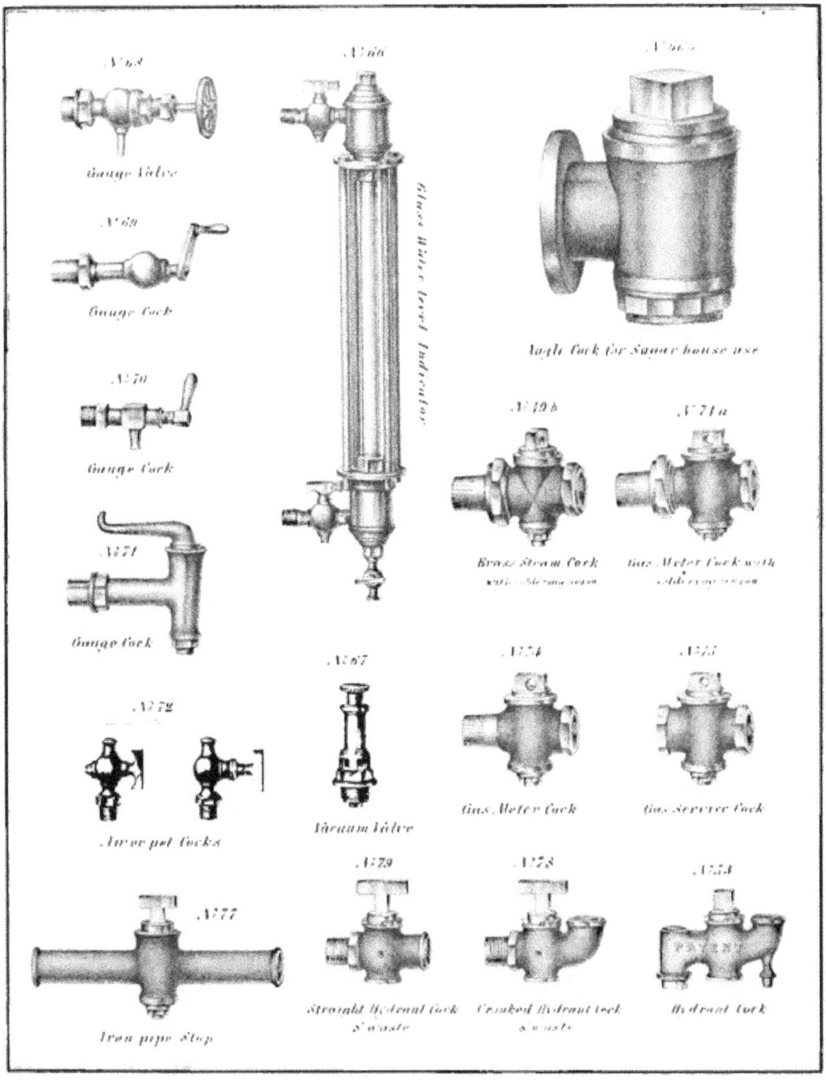
Gauge Valve
No 69
Gauge Cock
No 70
Gauge Cock
No 71
Gauge Cock
No 72
No 77
Iron pipe Stop
No 66
Glass Water level Indicator
No 67
Vacuum Valve
Angle Cock for Sugar house use
Brass Steam Cock
Gas Meter Cock with
Gas Meter Cock
Gas Service Cock
No 79
Straight Hydrant Cock
& waste
No 78
Crooked Hydrant Cock
& waste
Hydrant Cock

MORRIS, TASKER & Co.'s ILLUSTRATED CATALOGUE.

Plate 8

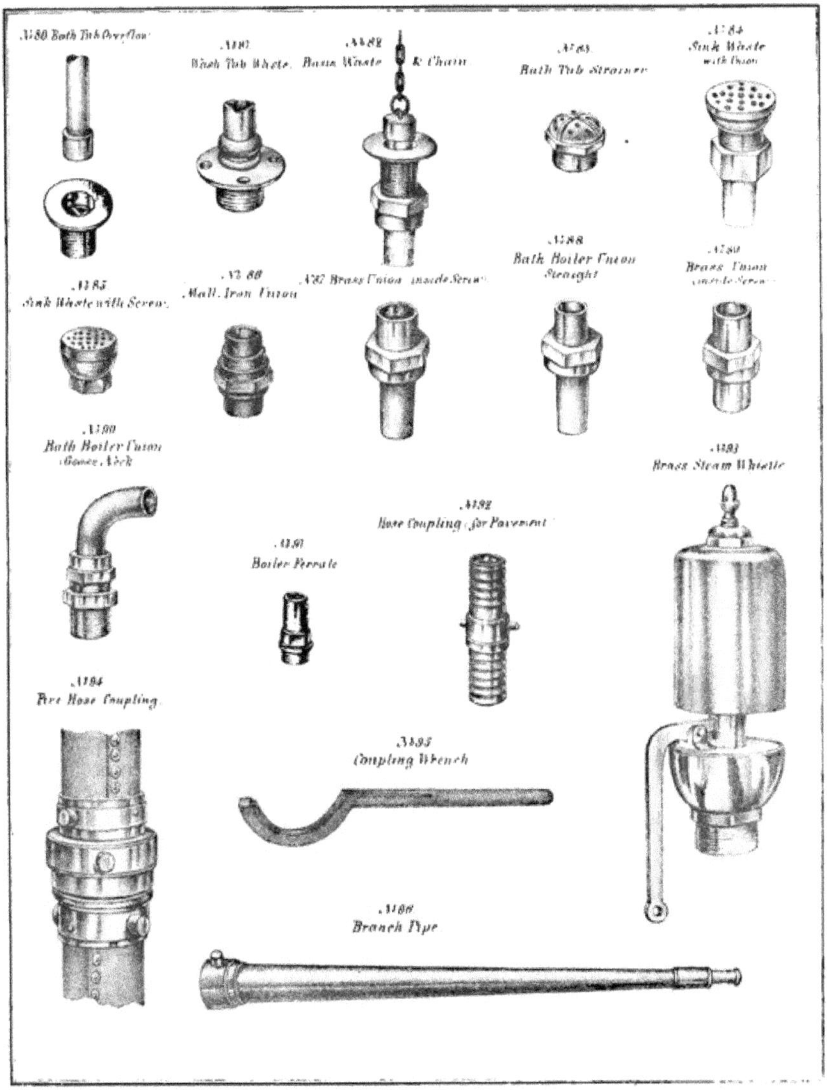

Plate 9

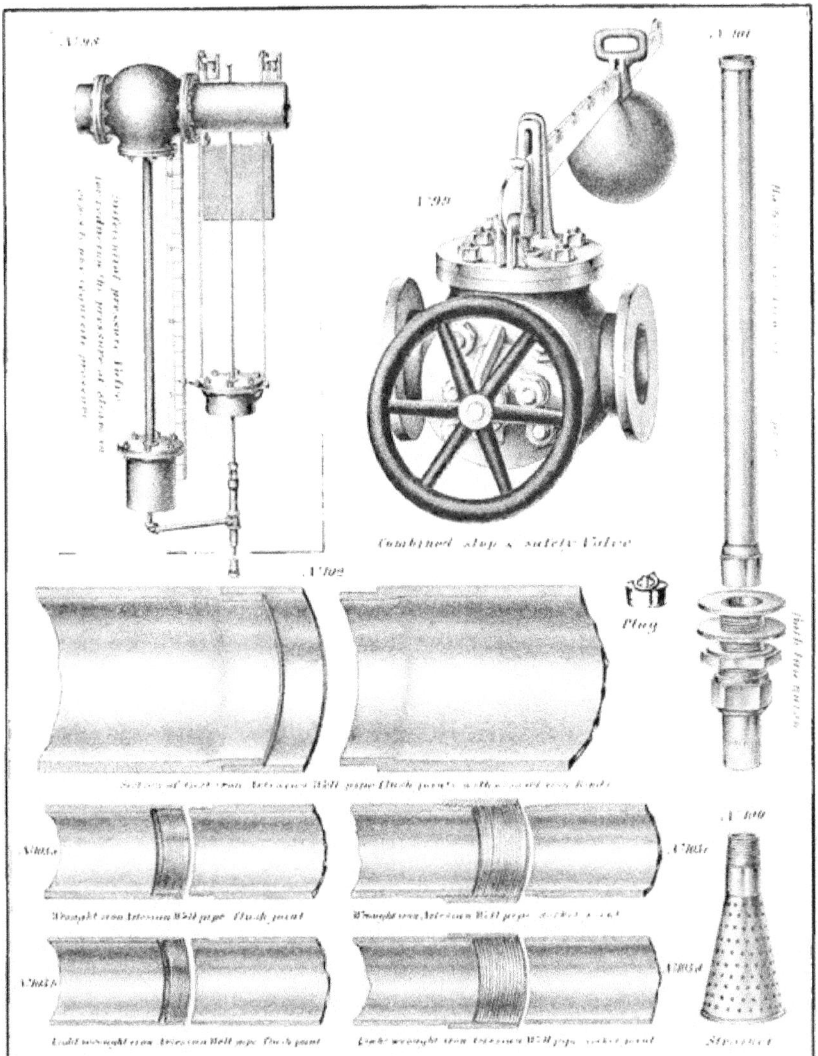

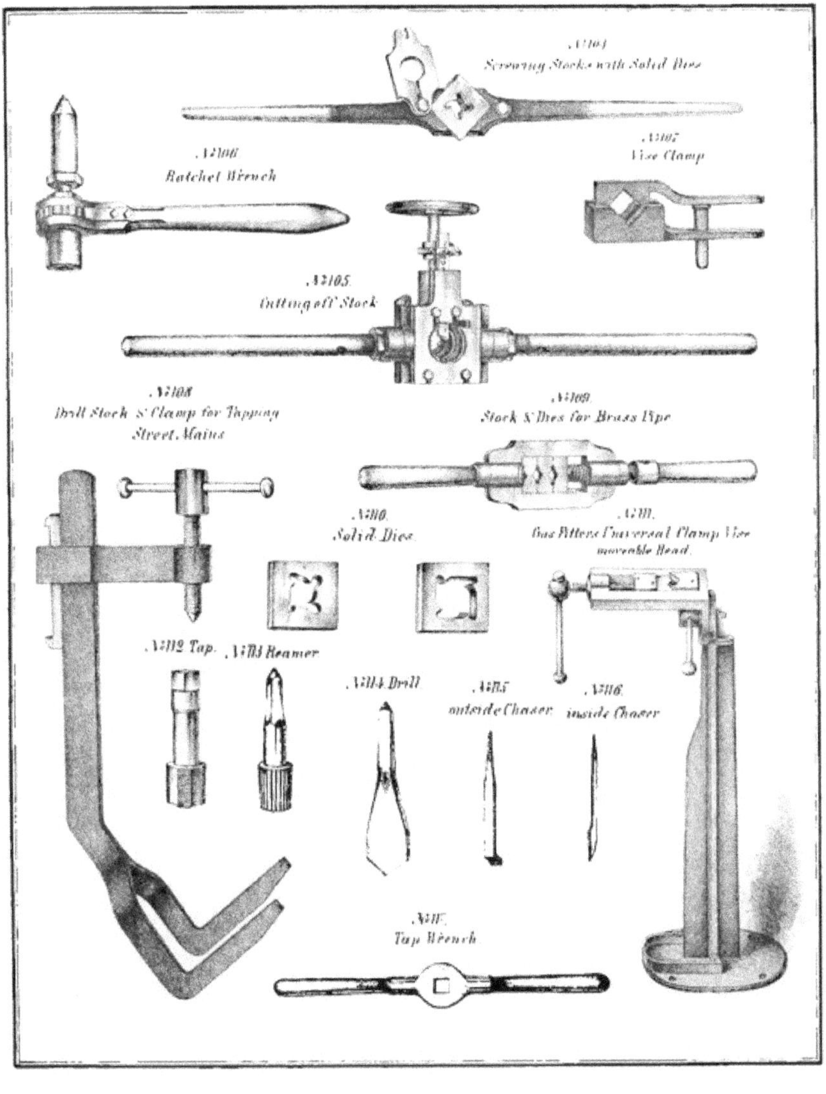

No. 104
Screwing Stocks with Solid Dies
No. 106
Ratchet Wrench
No. 107
Vise Clamp
No. 105
Cutting off Stock
No. 108
Drill Stock & Clamp for Tapping Street Mains
No. 109
Stock & Dies for Brass Pipe
No. 110
Solid Dies
No. 111
Gas Fitters Universal Clamp Vise moveable Head
No. 112 Tap
No. 113 Reamer
No. 114 Drill
No. 115
outside Chaser
No. 116
inside Chaser
No. 117
Tap Wrench

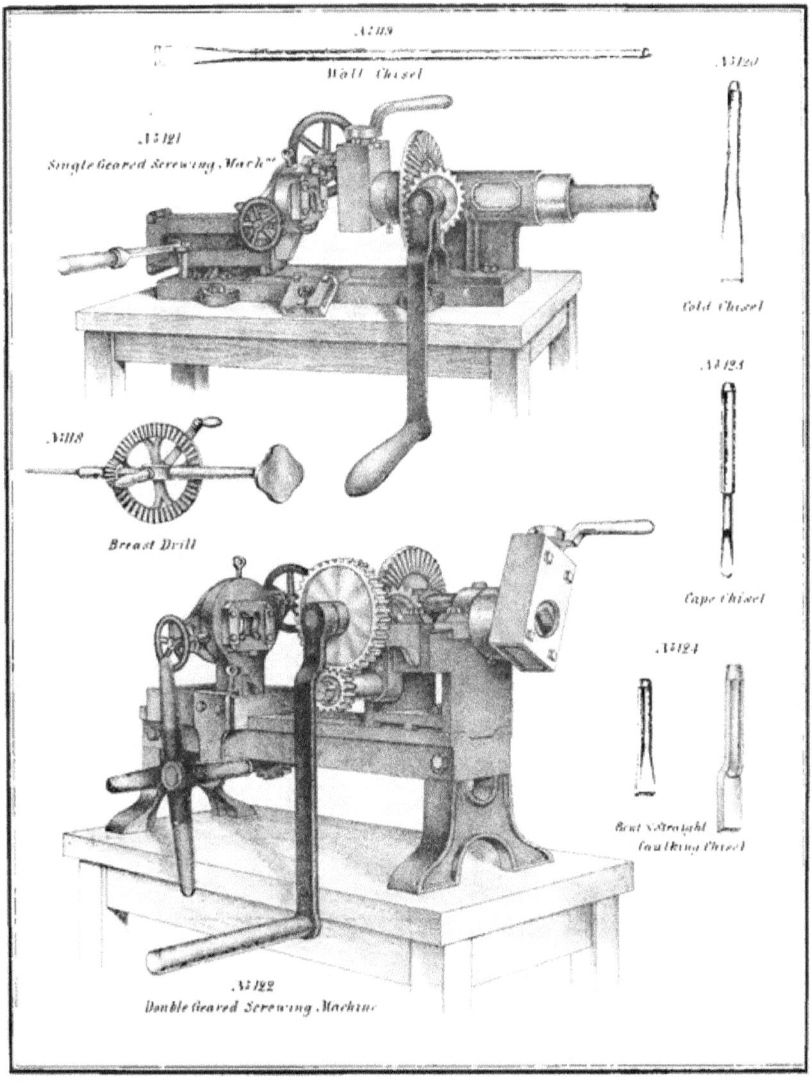
No 119
Wall Chisel
No 120
Cold Chisel
No 121
Single Geared Screwing Machne
No 118
Breast Drill
No 123
Cape Chisel
No 124
Bent & Straight
Caulking Chisel
No 122
Double Geared Screwing Machine

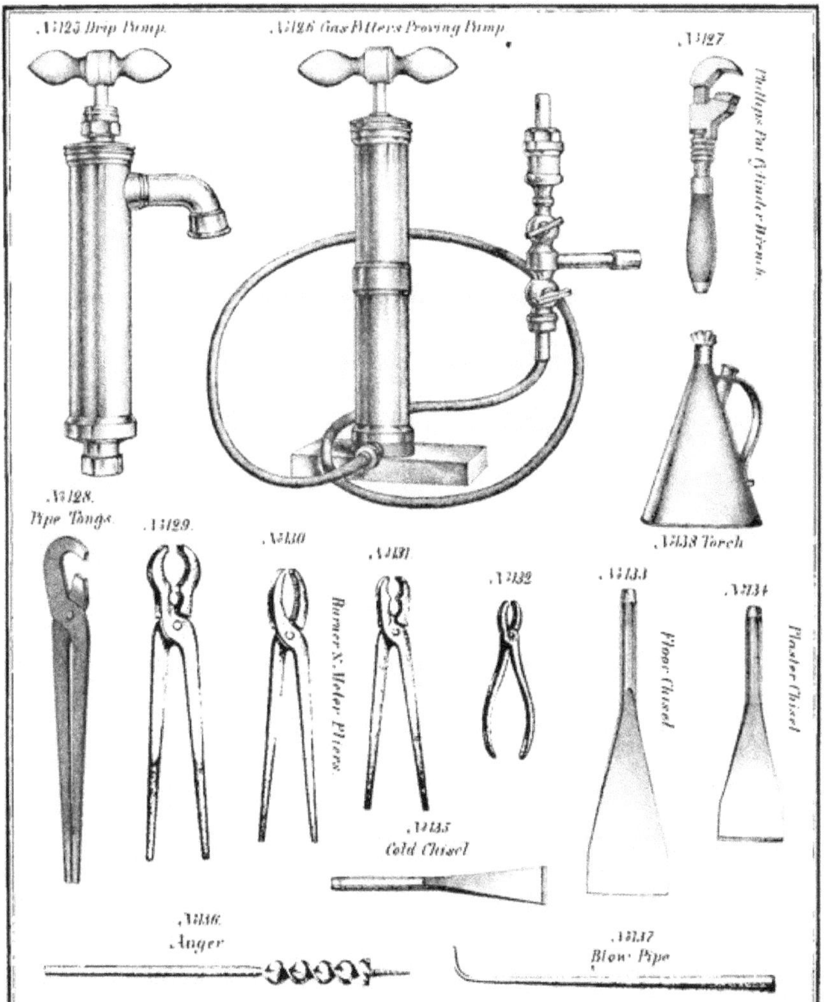
No. 125 Drip Pump.
No. 126 Gas Fitters Proving Pump
No. 127
Phillips Pat. Cylinder Wrench.
No. 128.
Pipe Tongs.
No. 129.
No. 130
Burner & Meter Pliers.
No. 131
No. 132
No. 133
Floor Chisel
No. 134
Plaster Chisel
No. 138 Torch
No. 135
Cold Chisel
No. 136.
Auger
No. 137
Blow Pipe

MORRIS, TASKER & Co's ILLUSTRATED CATALOGUE.

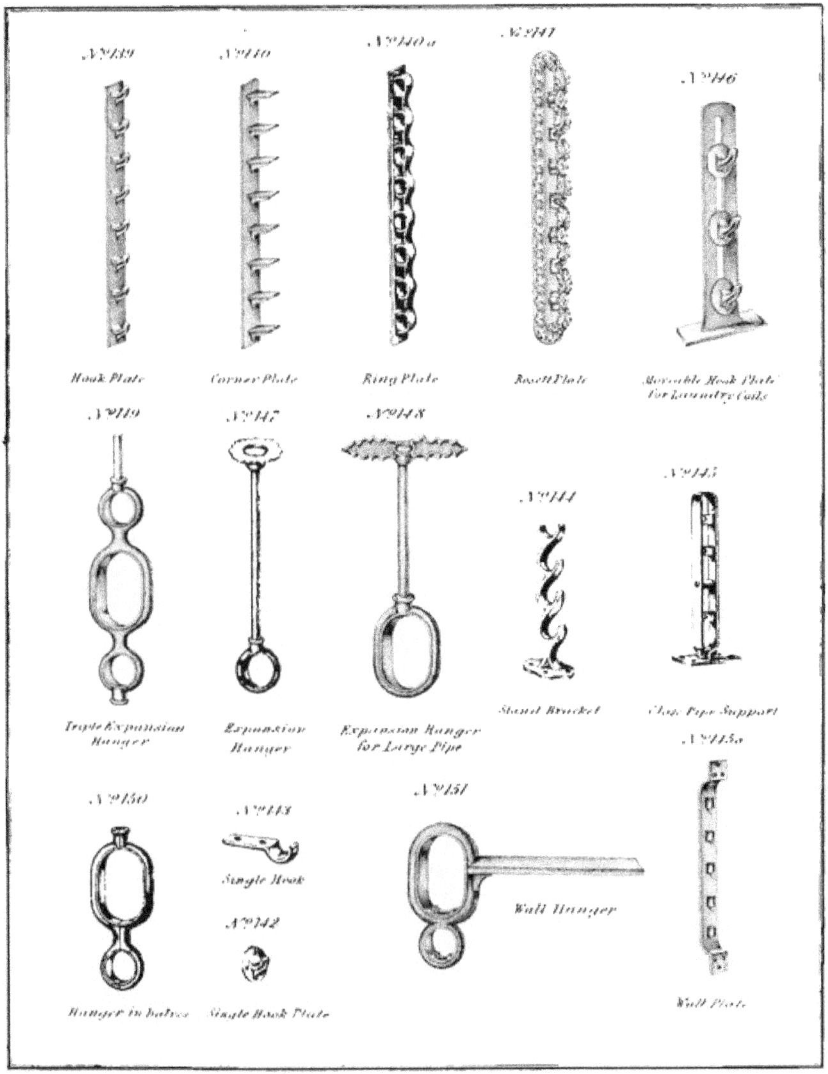

Plate 14

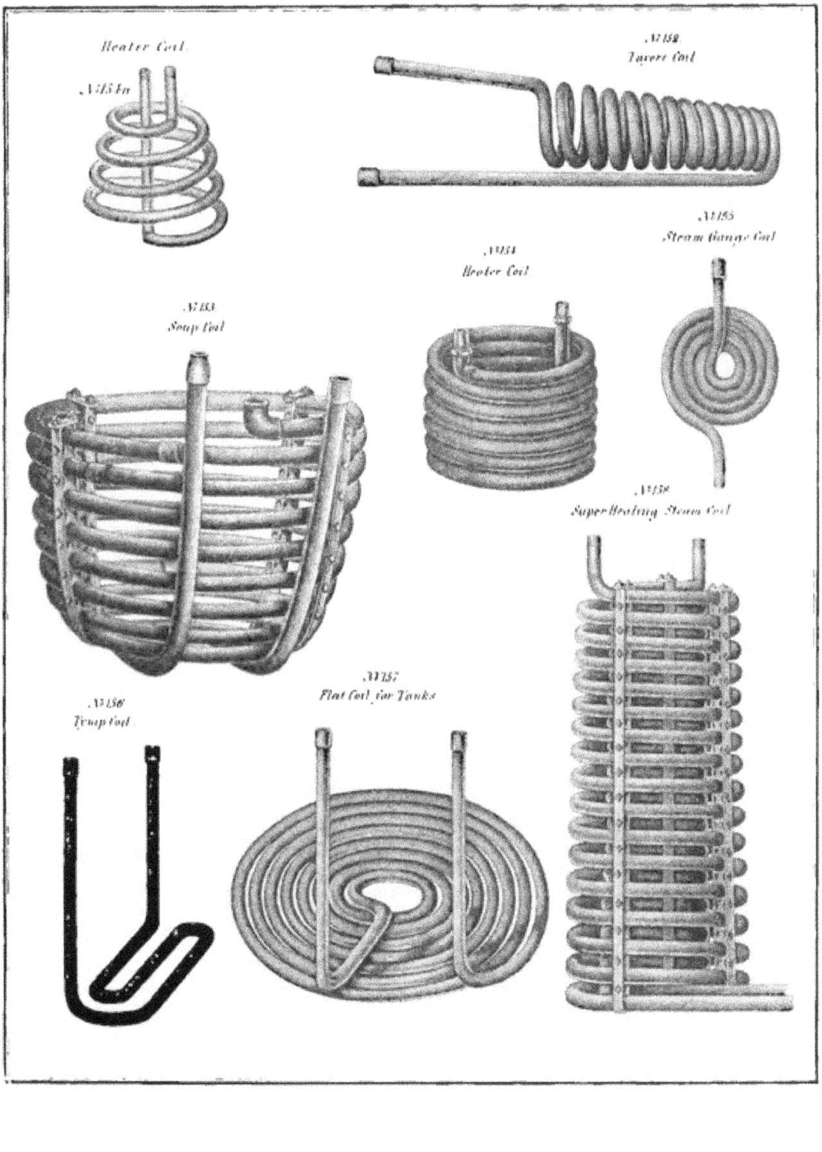
Heater Coil.
No 151a
No 152.
Tuyere Coil
No 155
Steam Gauge Coil
No 154
Heater Coil
No 153
Soup Coil
No 158
Super Heating Steam Coil
No 157
Flat Coil for Tanks
No 156
Trap Coil.

Plate 15

N° 159

Radiator for Walls with Manifolds.

N° 160

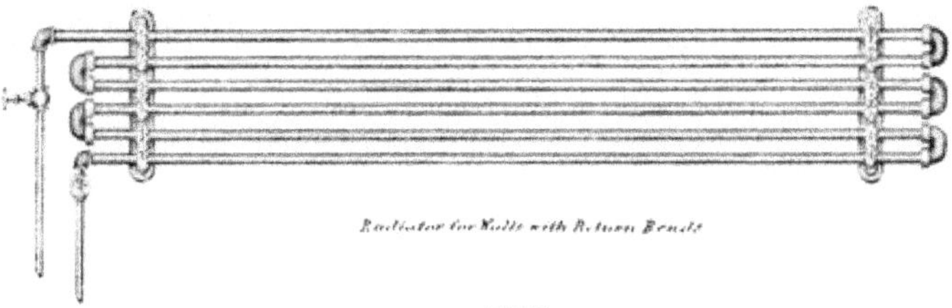

Radiator for Walls with Return Bends.

N° 161

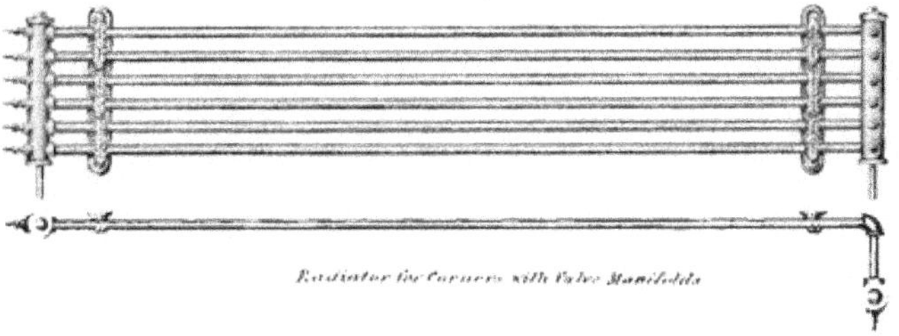

Radiator for Corners with Valve Manifolds.

N° 162

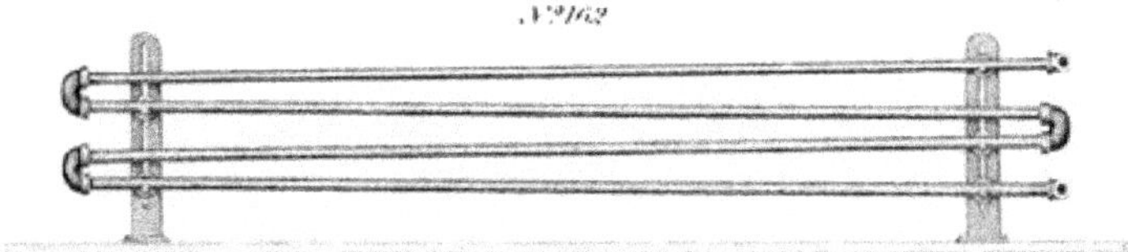

Radiator for Drying Closets with Return Bends.

Plate 16

No 163

Ornamental Radiator for Walls

No 164

Cast Iron Pedestal Radiator

No. 166.

Screen for covering Box Coils.

No. 165.

Box Coil Radiator.

Plate Ten

No 168

Screens for Covering Box Coils

No 167

No. 169 a

Screen for Recess Radiator

No. 169

Ornamental Screen for Box Coil Radiator

Nº 167

Double acting Force & Lift Pump

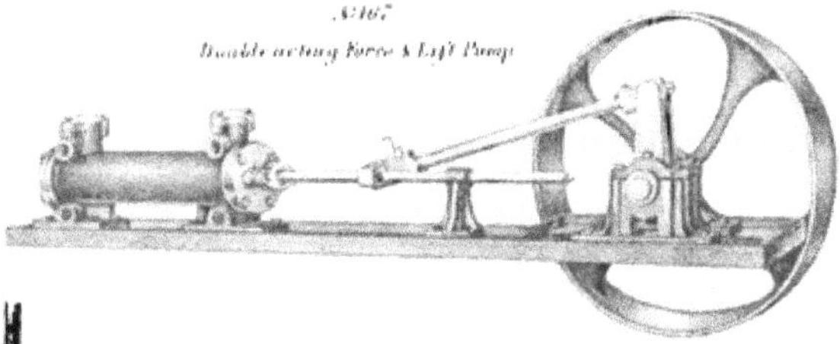

Nº 168

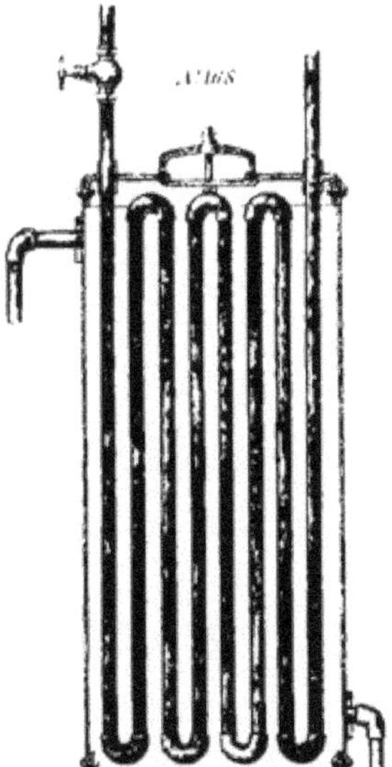

Hot Water Tank with Coil

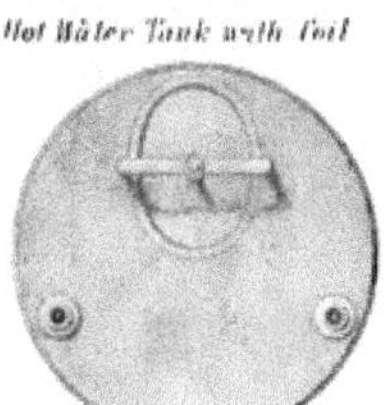

Nº 170

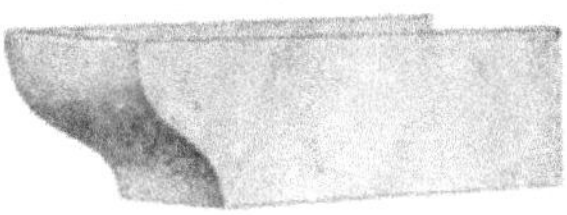

Battlement Bracket

Nº 170½

Battlement Bracket

Nº 169

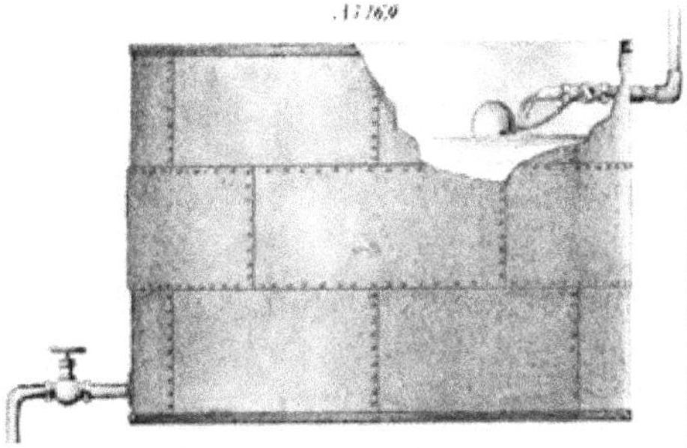

Cold Water Tank with Ball Cock.

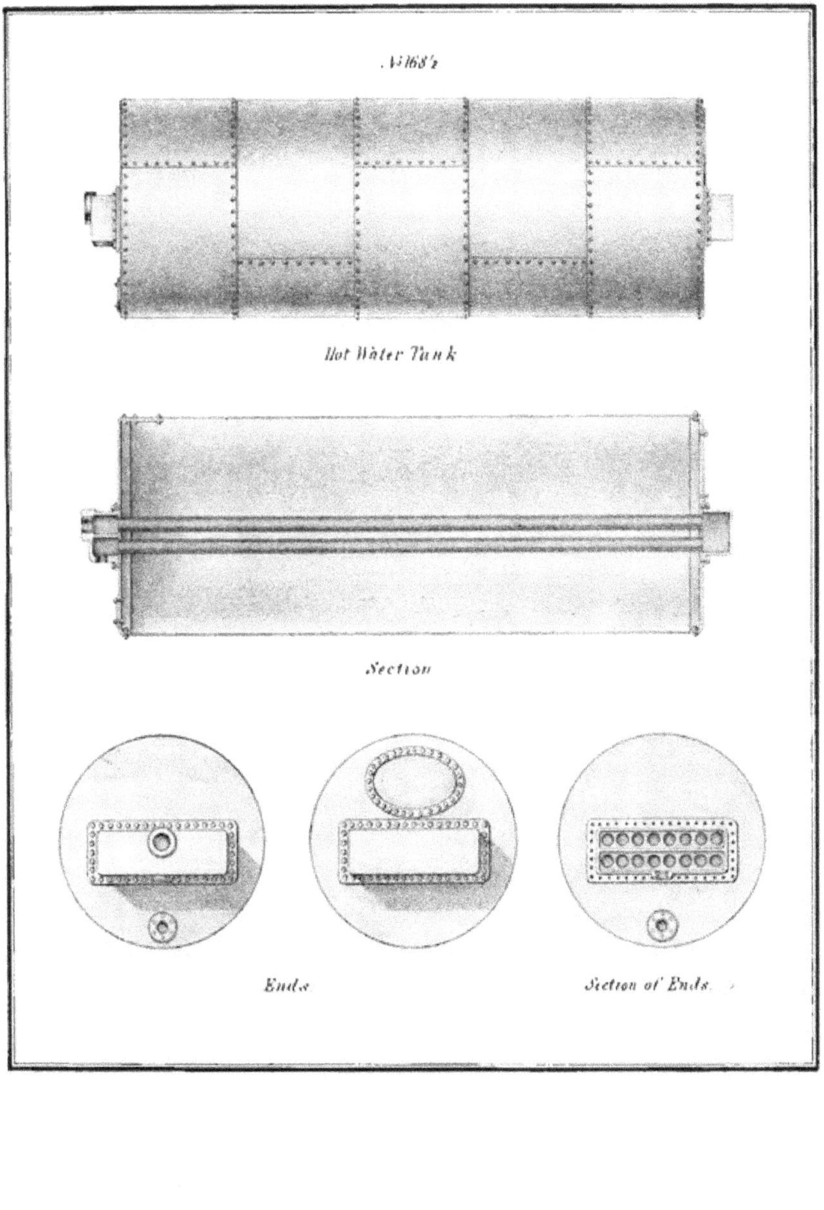
No 168½
Hot Water Tank
Section
Ends
Section of Ends.

No. 171

Horizontal Tubular Boiler

with Fire Box

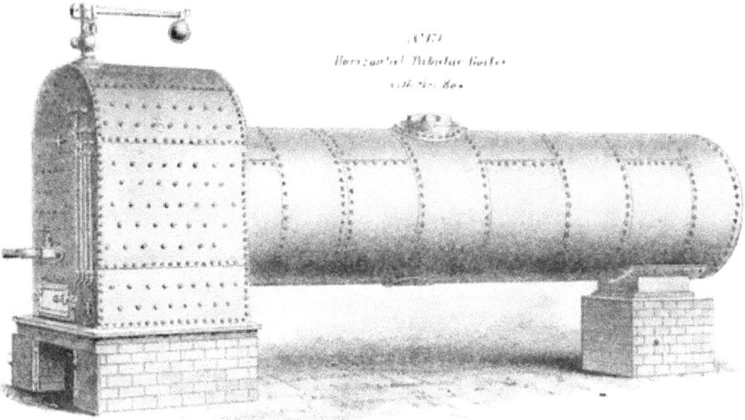

No. 172

Vertical Tubular Boiler

No. 173

Circulating Hot Water Boiler

No. 174

Hot Water Back

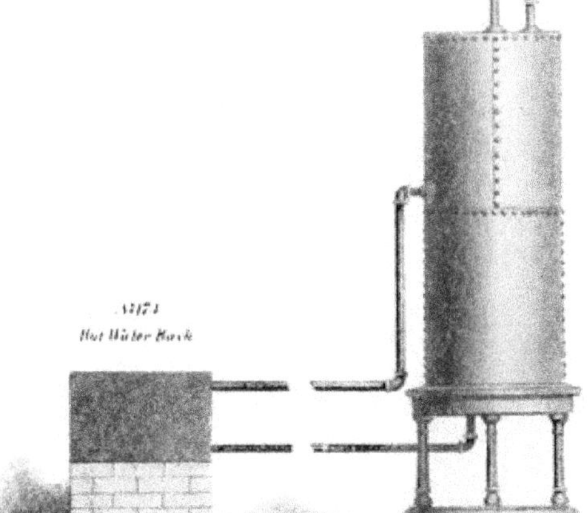

Plate 20

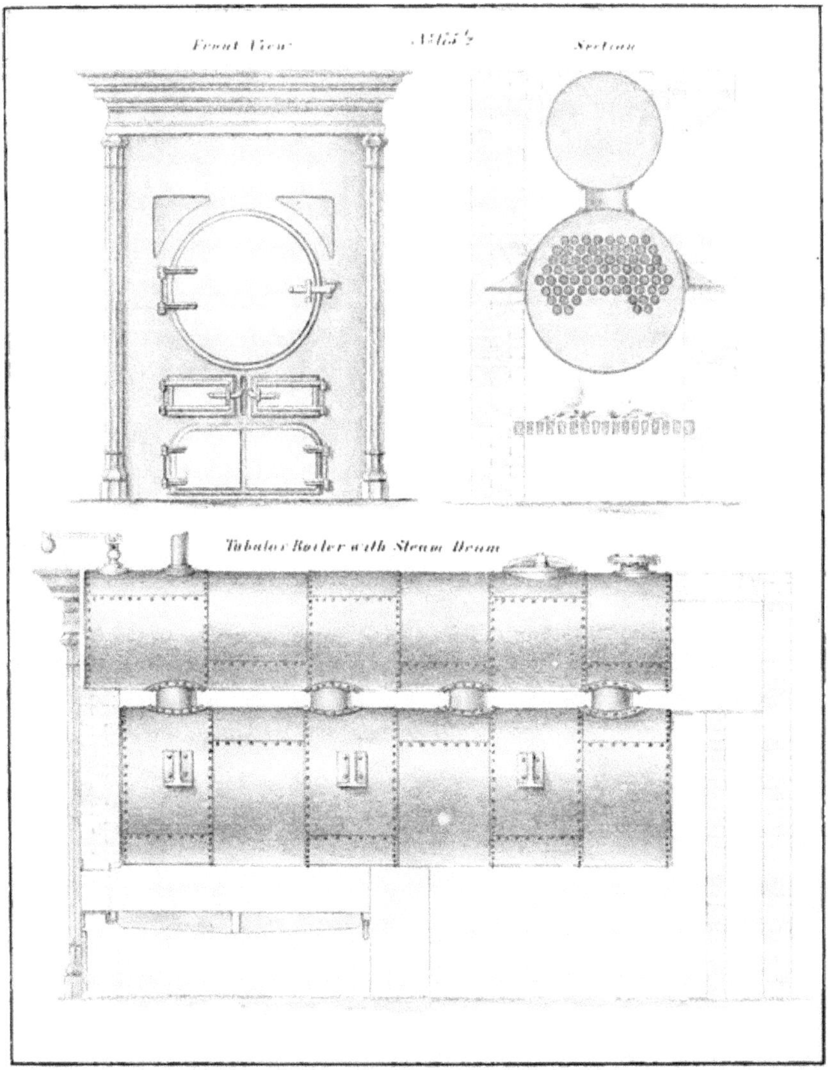

Artesian or other Well Pump

No 177

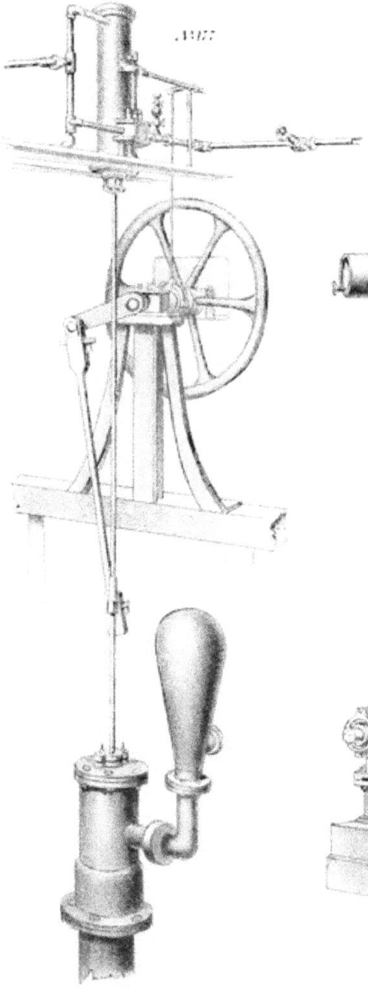

No 178

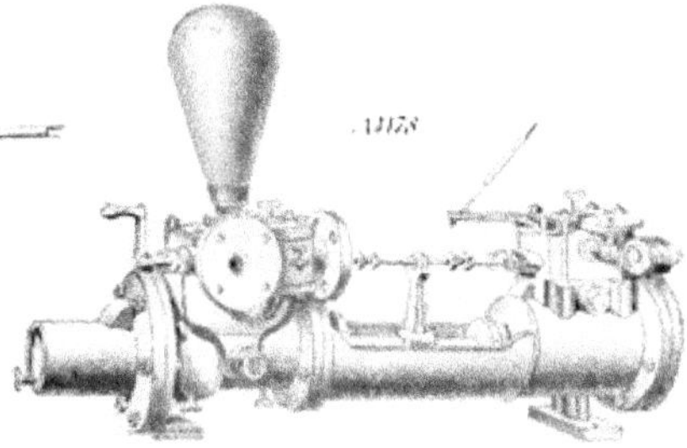

Worthington Pump

No 179

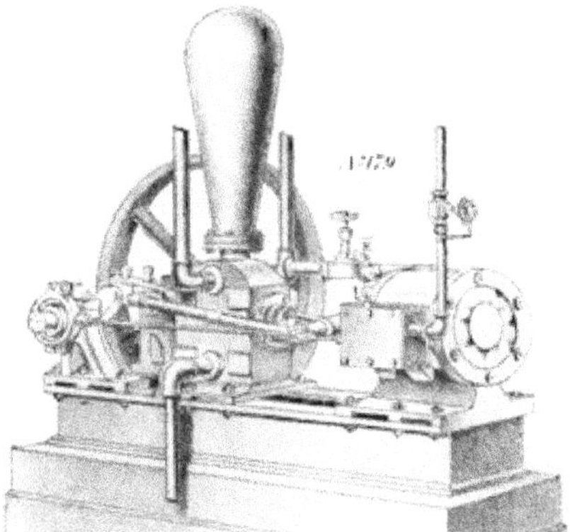

Woodward Pump

No 185

Ventilator Fronts.

No 186

Backs & Jambs for Fire places.

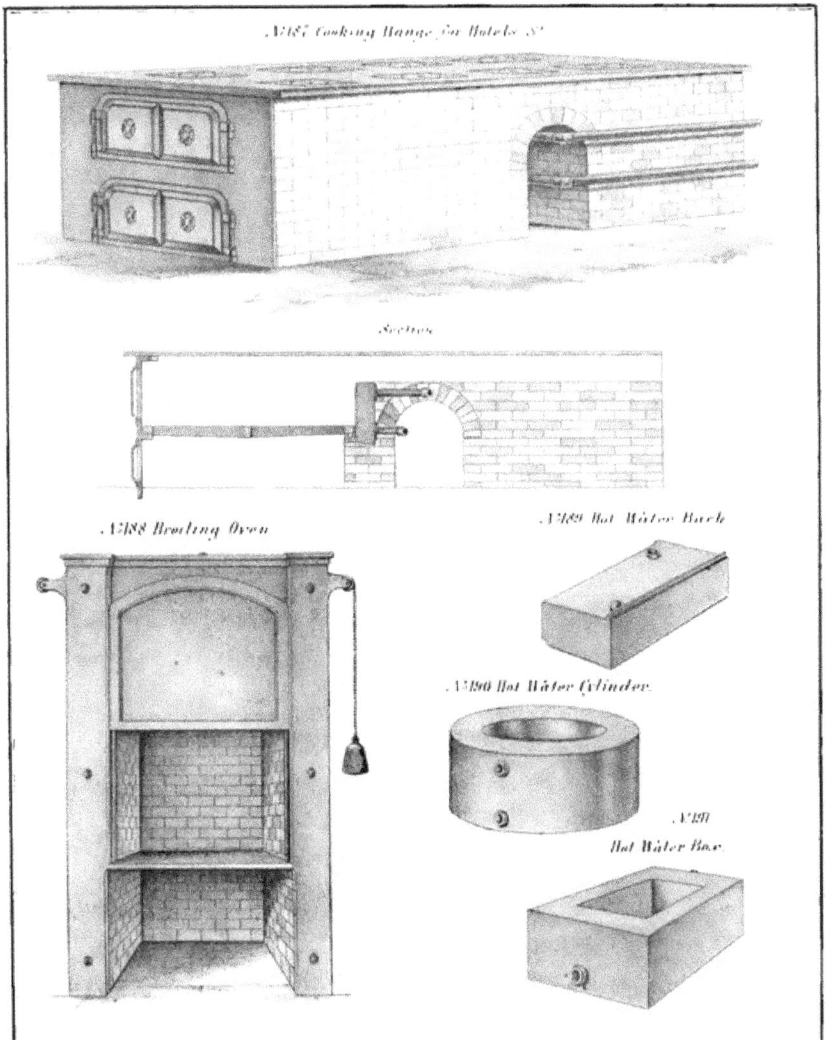
No. 187 Cooking Range for Hotels &c.
Section
No. 188 Broiling Oven
No. 189 Hot Water Back
No. 190 Hot Water Cylinder.
No. 191
Hot Water Box.

No. 182

Rotary Roaster

No. 183

Rousing Tub

No. 184

Boiling Tub

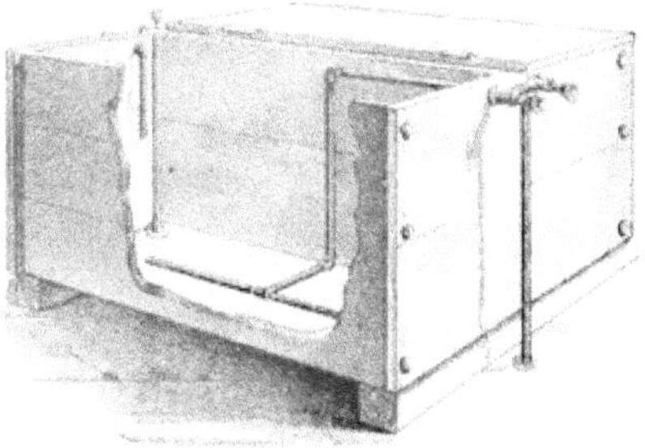

No. 185

Round & Oval Tin Steamers

No. 186

Mash Tubs

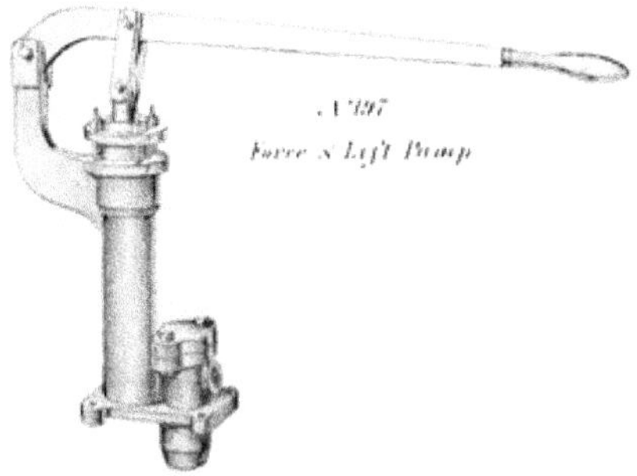

No. 197
Force & Lift Pump

No. 198 Steam Trap

No. 193 Lard Kettle

No. 200
Clothes Wringer

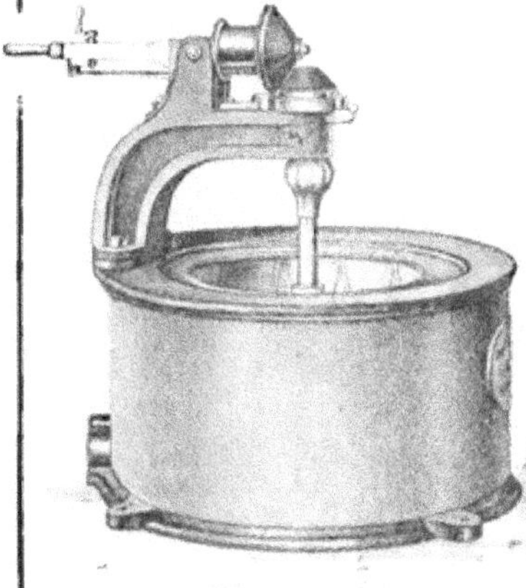

No. 201

Double Steam Kettle

No. 202

Shaker Washing Machines

No. 203

Drying Closet

No. 203 b

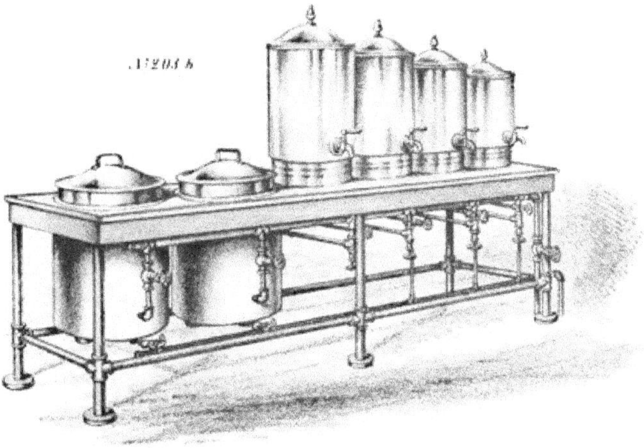

Steam Table for warming Soup - Coffee - Tea & Milk.

No. 203 a.

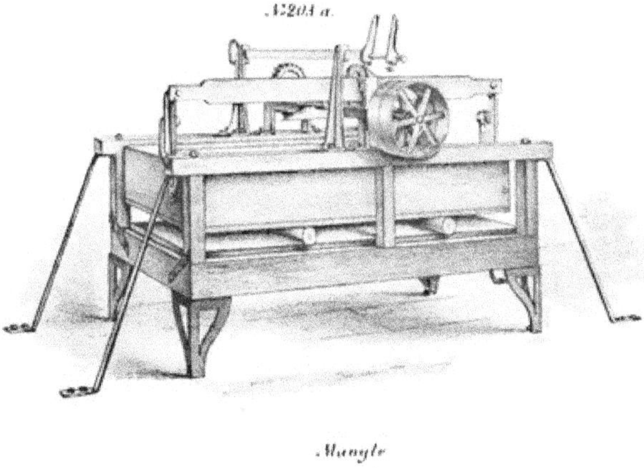

Mangle

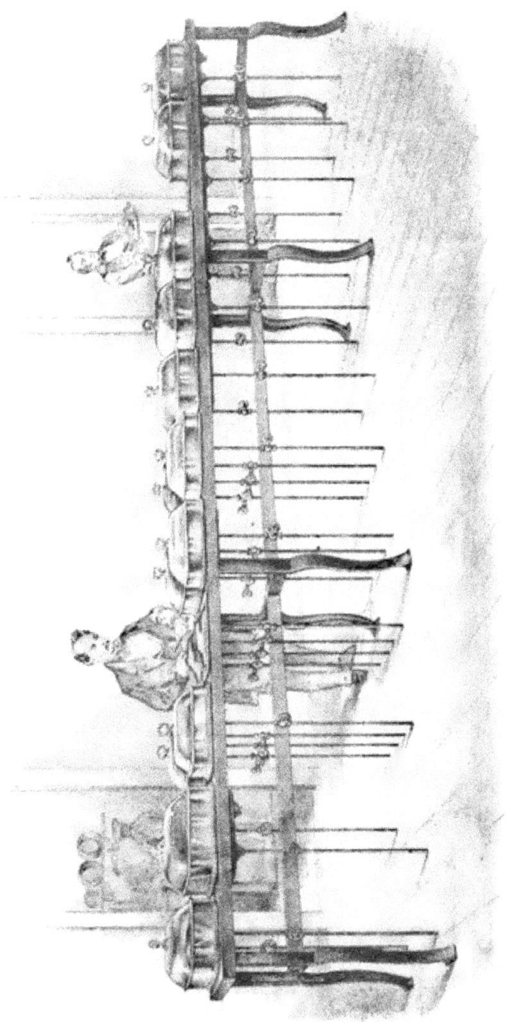

No 201
Steam Carving Table and Dishes
with [illegible]

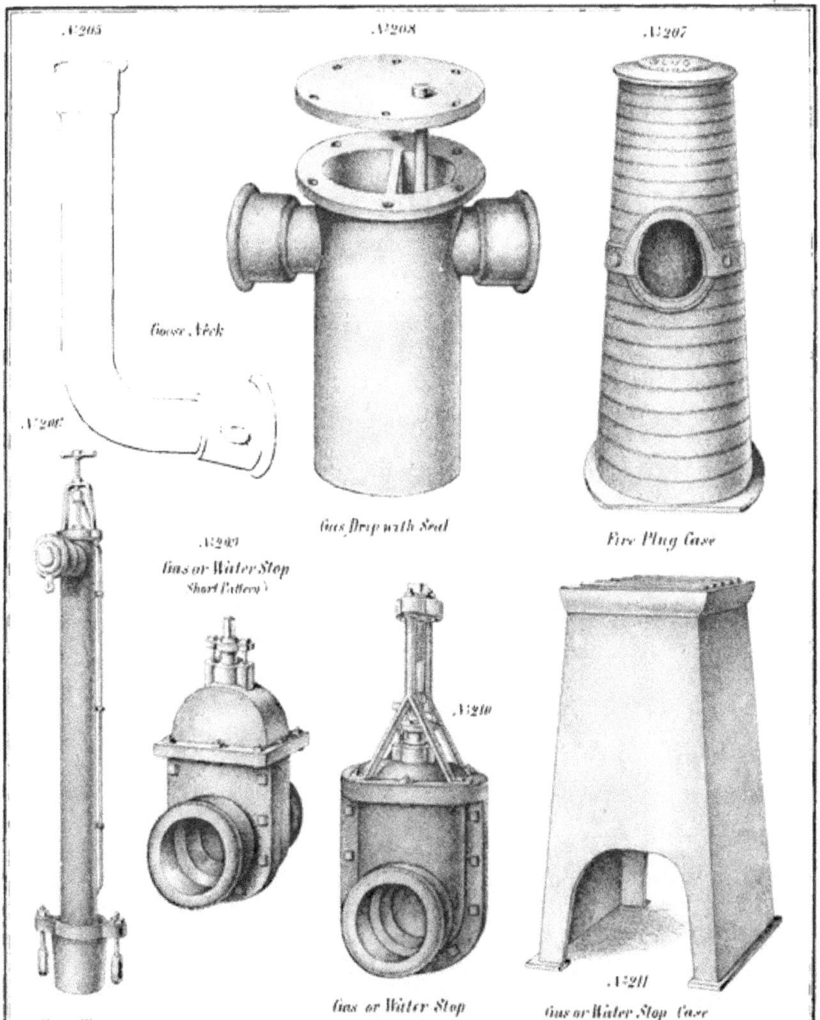
No. 205
Goose Neck
No. 208
Gas Drip with Seal
No. 207
Fire Plug Case
No. 206
Fire Plug
No. 209
Gas or Water Stop
(Short Pattern)
No. 210
Gas or Water Stop
for Streets
No. 211
Gas or Water Stop Case

Plate 31a

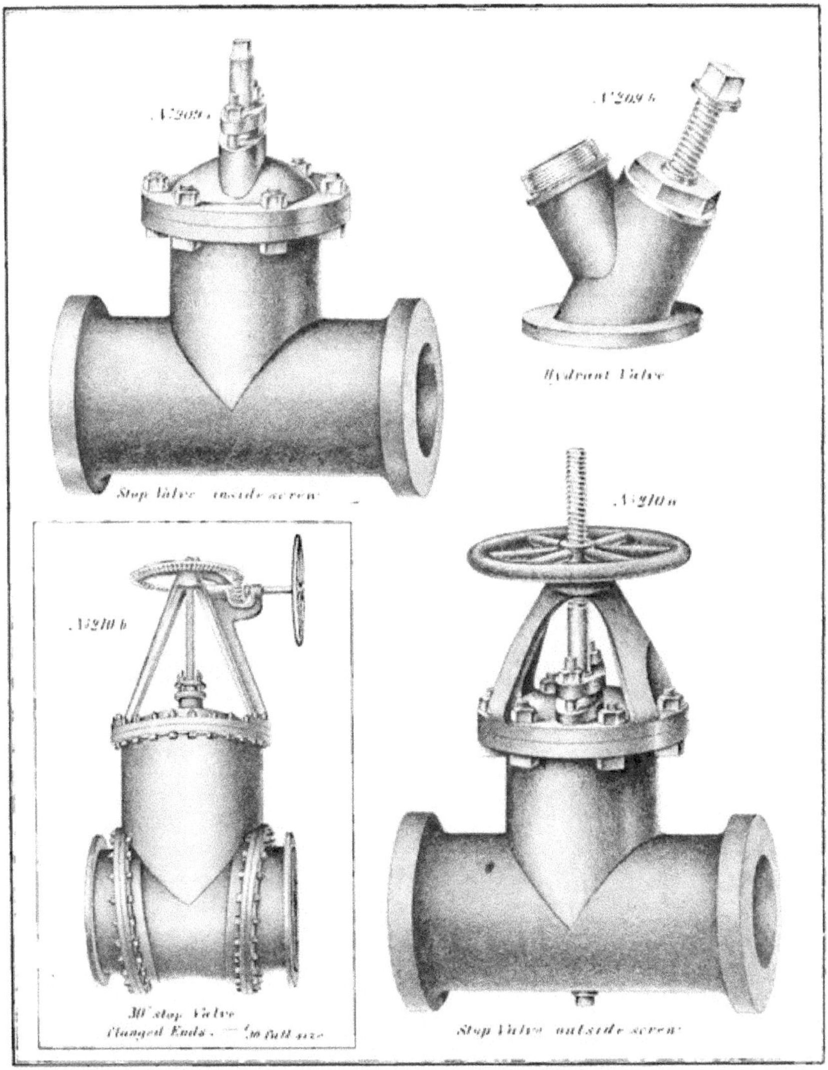
No. 209 a
No. 209 b
Hydrant Valve
Stop Valve inside screw
No. 210 a
No. 210 b
30" stop Valve
Flanged Ends. 1/10 full size
Stop Valve outside screw

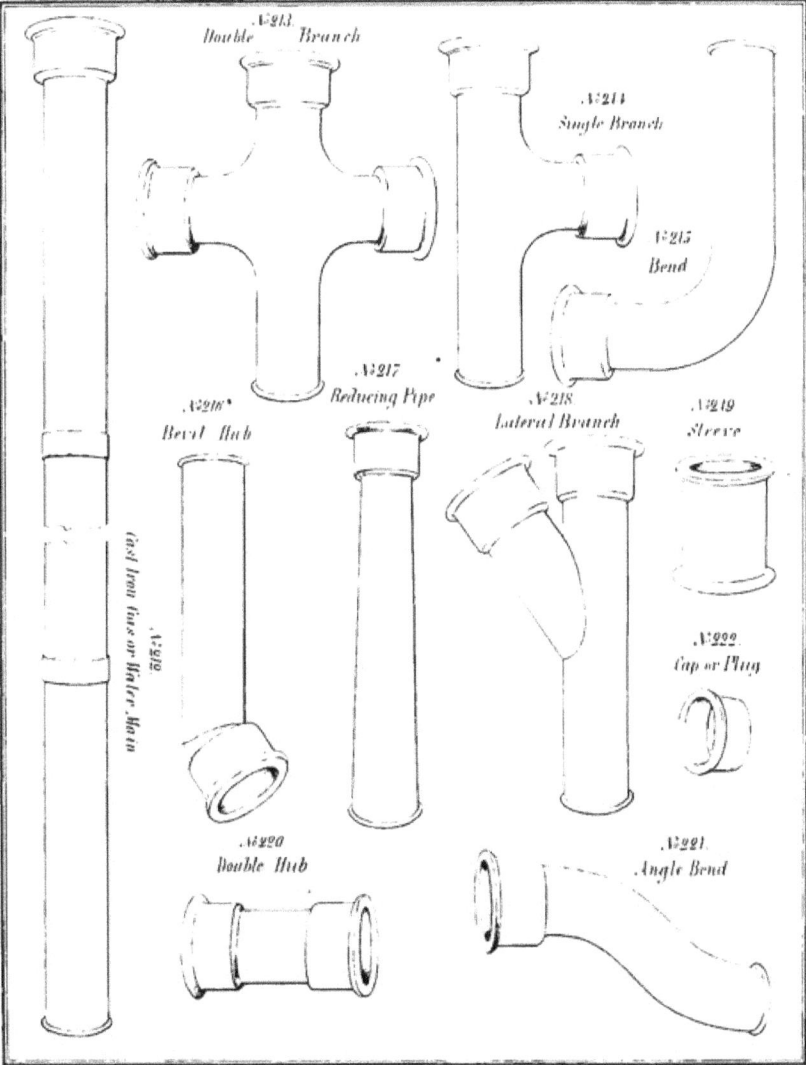

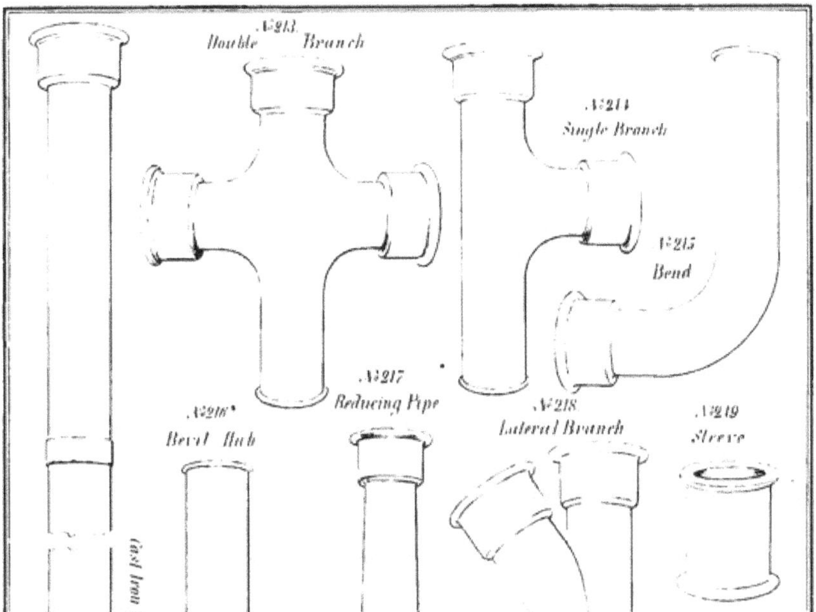
No. 213.
Double Branch
No. 214
Single Branch
No. 215
Bend
No. 212.
Cast Iron Gas or Water Main
No. 216*
Bevel Hub
No. 217
Reducing Pipe
No. 218.
Lateral Branch
No. 219
Sleeve
No. 222.
Cap or Plug
No. 220
Double Hub
No. 221.
Angle Bend

No. 223 Water Closet Arrangements

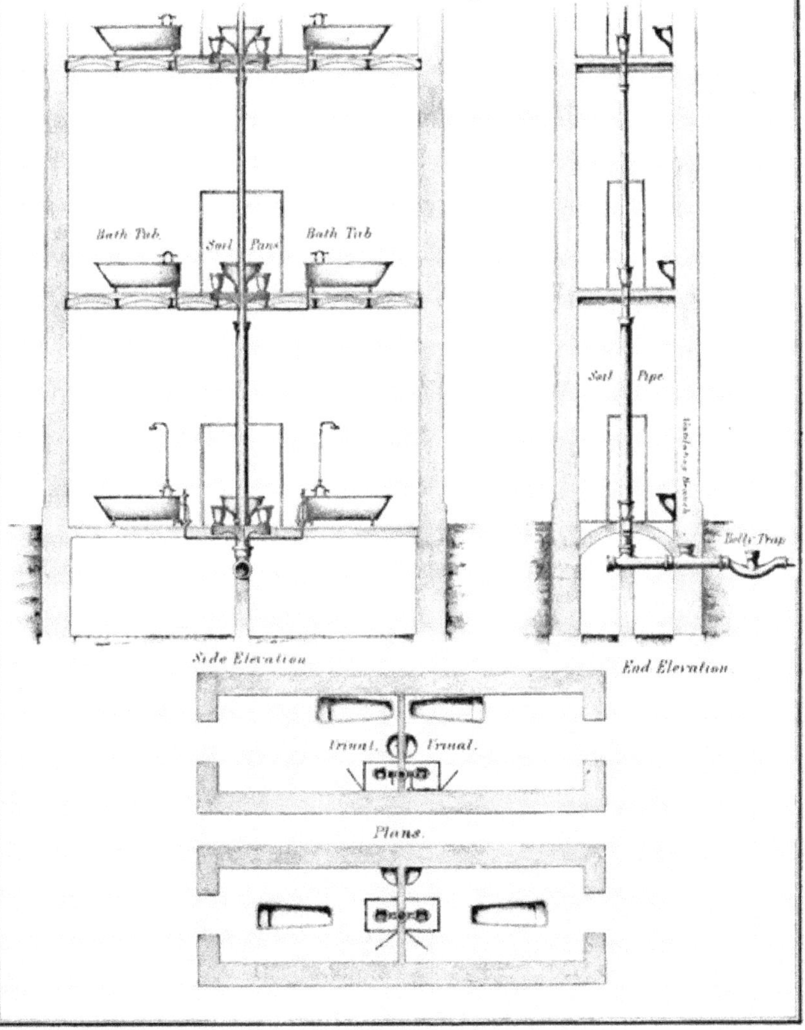

No. 223½

Plan of Water Closet Arrangements

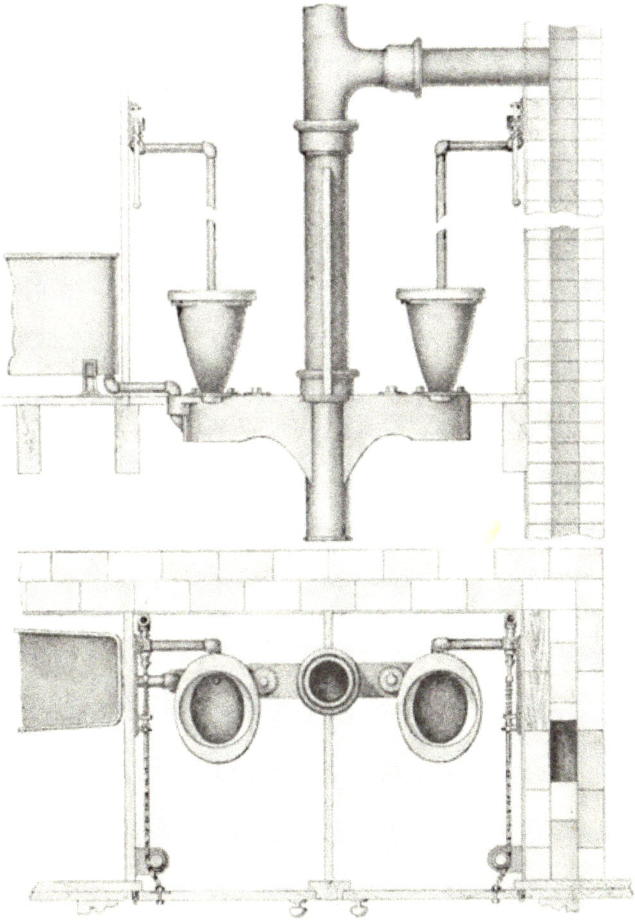

No. 224 Single Soil Branch

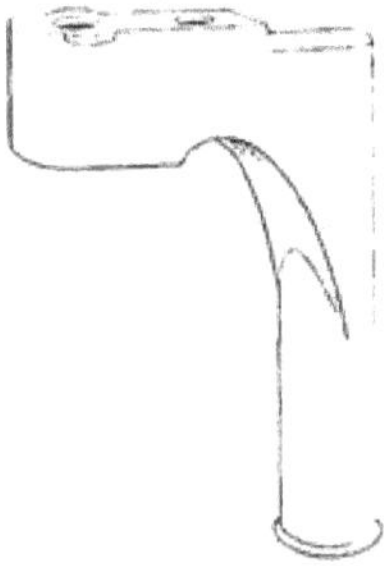

No. 225 Single Soil Branch

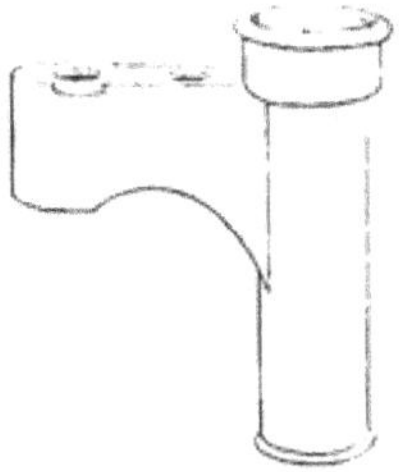

No. 226 Double Soil Branch.

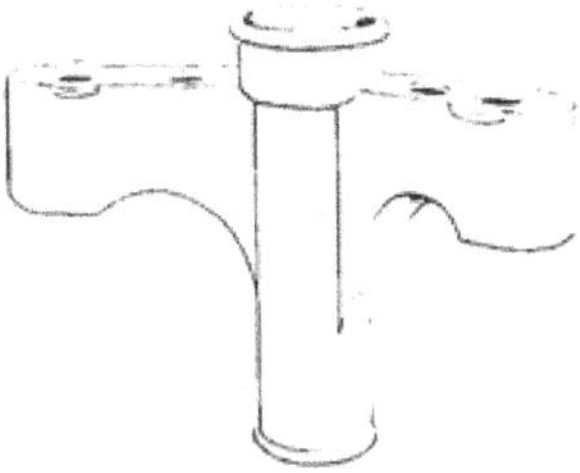

No. 228.
Bath Tub with Supply Waste & Overflow Combined

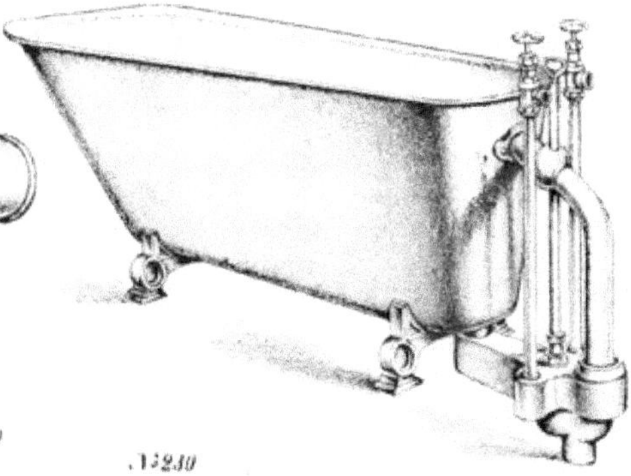

No. 227
Reservoir Valve
with Strainer

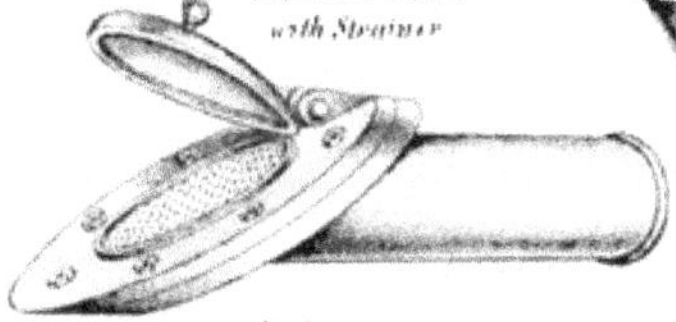

No. 229
Wash Basin with Feet & Waste & Cocks for
Hot & Cold Water

No. 230

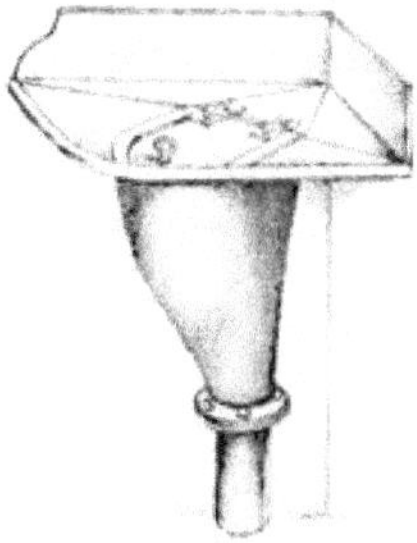

Slop Hopper with Lid

No. 231

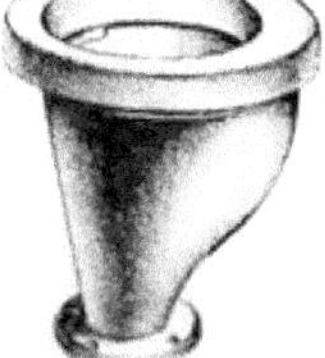

Soil Pan

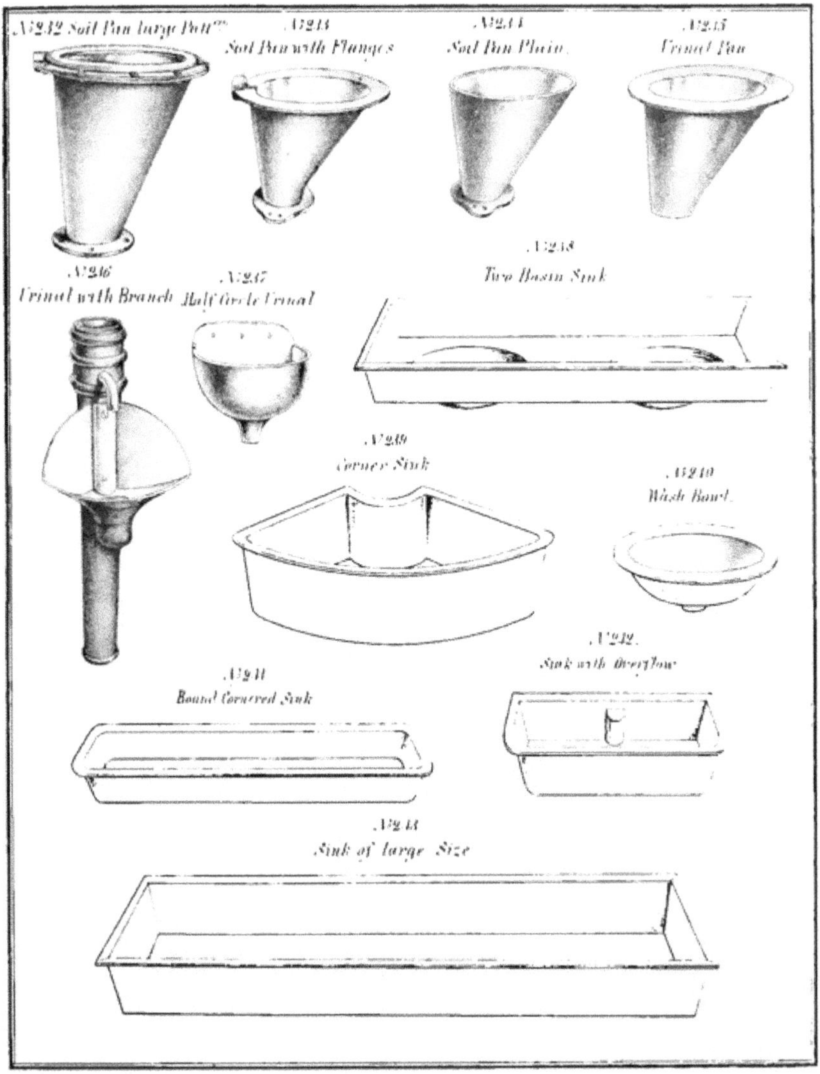

Nº 232 Soil Pan large Pattn
Nº 233
Soil Pan with Flanges
Nº 234
Soil Pan Plain.
Nº 235
Urinal Pan
Nº 238
Two Basin Sink
Nº 236
Urinal with Branch
Nº 237
Half Circle Urinal
Nº 239
Corner Sink
Nº 240
Wash Bowl.
Nº 242.
Sink with Overflow
Nº 241
Round Cornered Sink
Nº 243
Sink of large Size

MORRIS, TASKER & C°S ILLUSTRATED CATALOGUE.

No. 244.
Box Drain Trap.

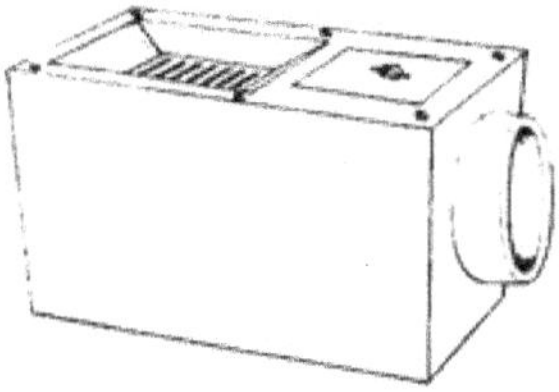

No. 245
Belly Trap

No. 246 Belly Trap with hand hole to clean

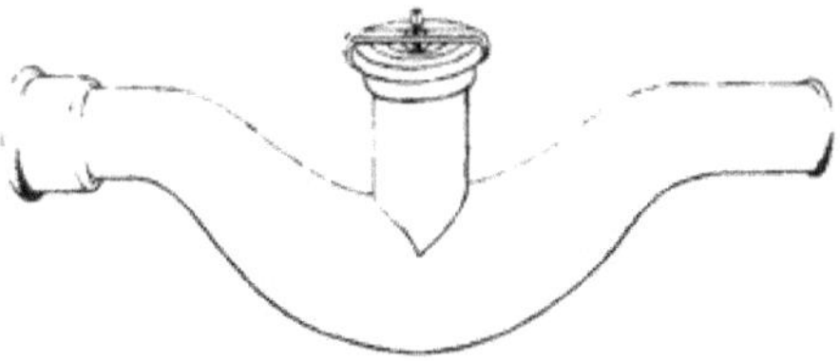

No. 247 Lateral Branch Box Trap.

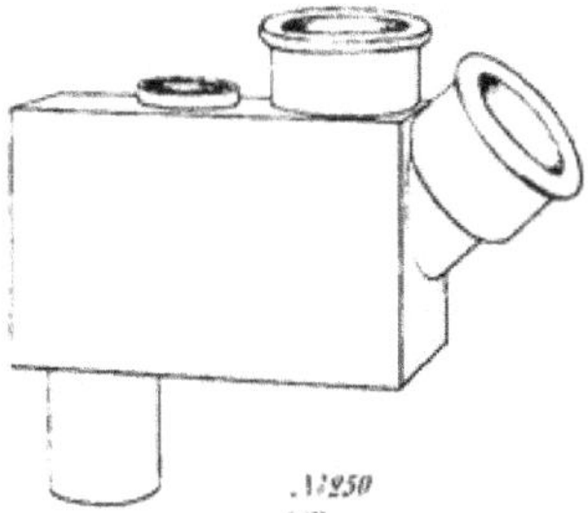

No. 248.
Right Angle Box Trap

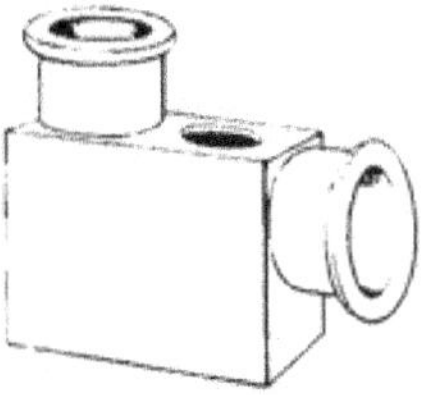

No. 249
Soil Pan Trap

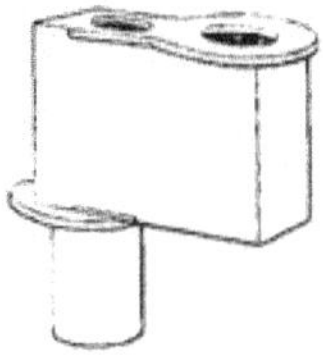

No. 250
S Trap

No. 251
Urinal Branch

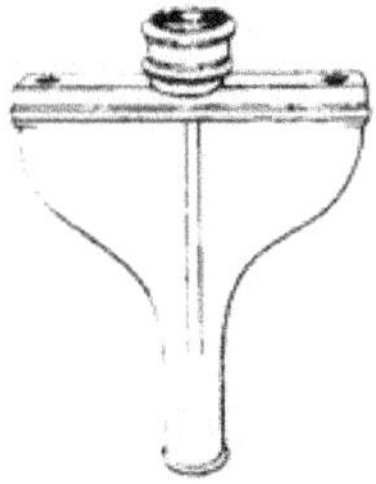

No. 252
Soil Pan Trap with Flanges

No. 251
U Trap

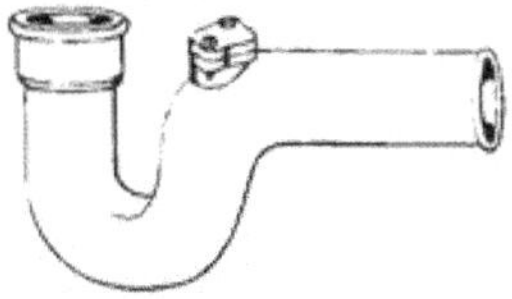

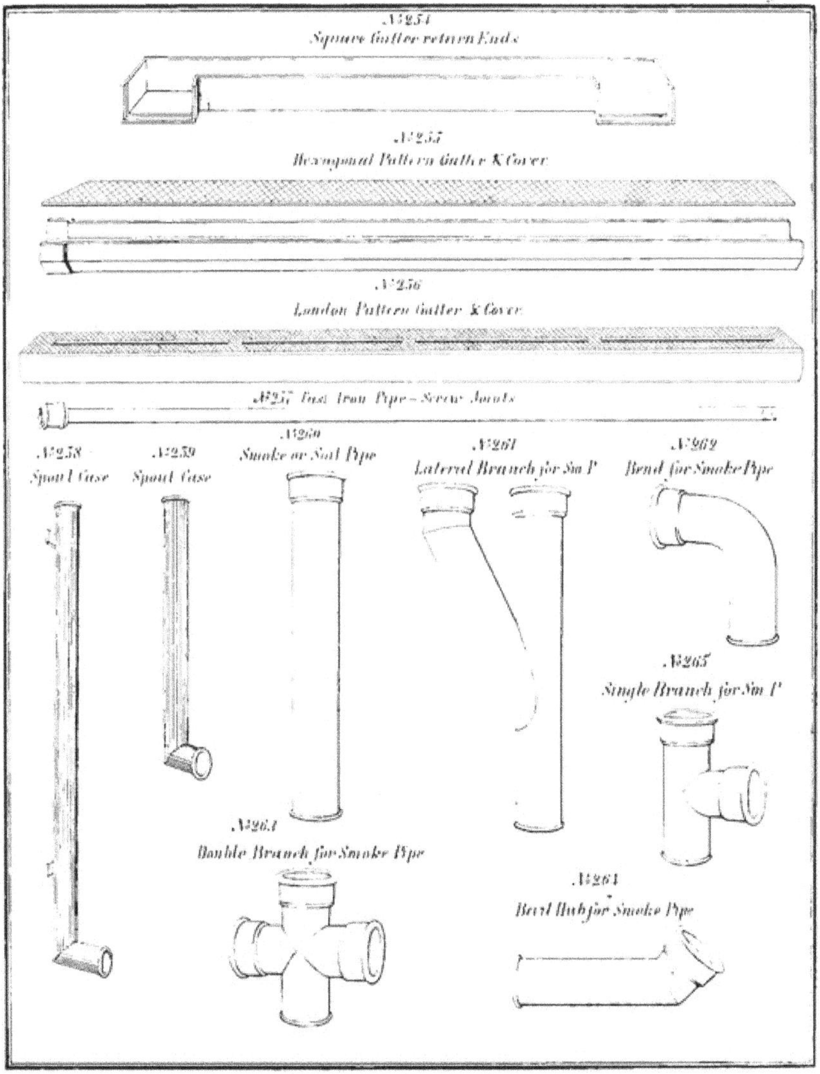
No 254
Square Gutter return Ends
No 255
Hexagonal Pattern Gutter & Cover
No 256
London Pattern Gutter & Cover
No 257 Cast Iron Pipe – Screw Joints
No 258
Spout Case
No 259
Spout Case
No 260
Smoke or Soil Pipe
No 261
Lateral Branch for Sm P
No 262
Bend for Smoke Pipe
No 265
Single Branch for Sm P
No 264
Double Branch for Smoke Pipe
No 261
Bevel Hub for Smoke Pipe

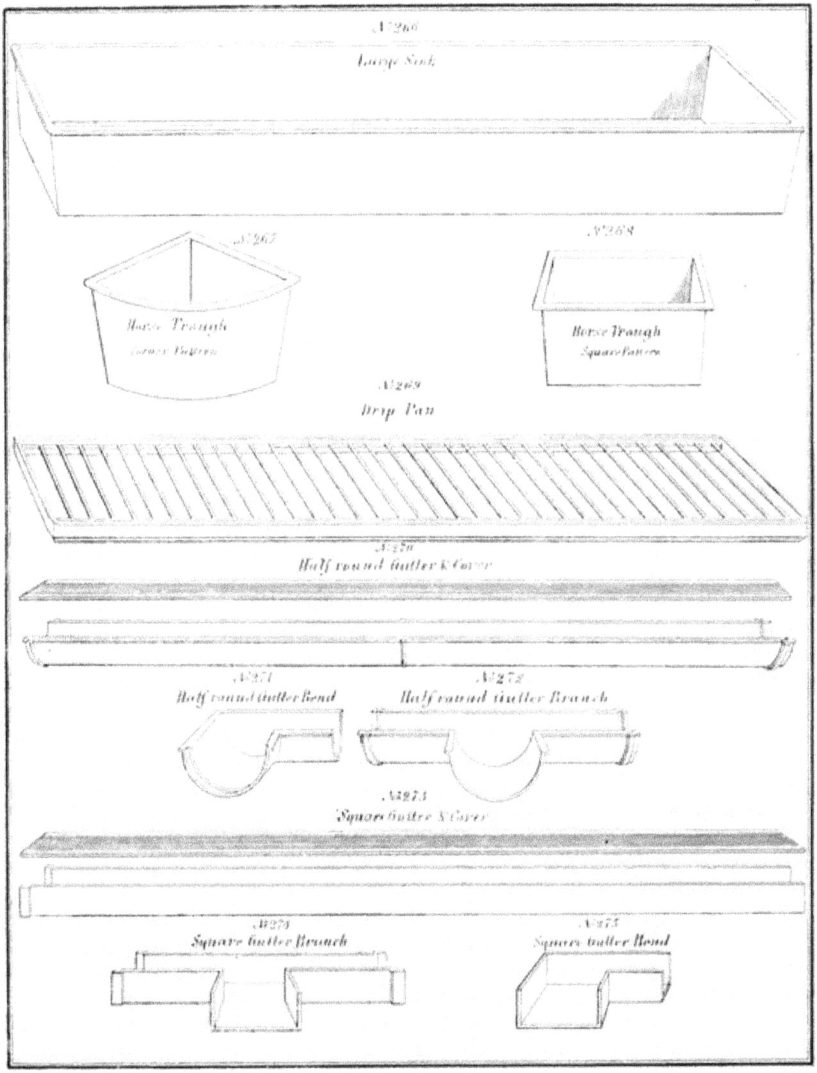
No 266
Large Sink
No 267
Horse Trough
Corner Pattern
No 268
Horse Trough
Square Pattern
No 269
Drip Pan
No 270
Half round Gutter & Cover
No 271
Half round Gutter Bend
No 272
Half round Gutter Branch
No 273
Square Gutter & Cover
No 274
Square Gutter Branch
No 275
Square Gutter Bend

No. 276
Tasker's Pat. non Freezing Hydrant

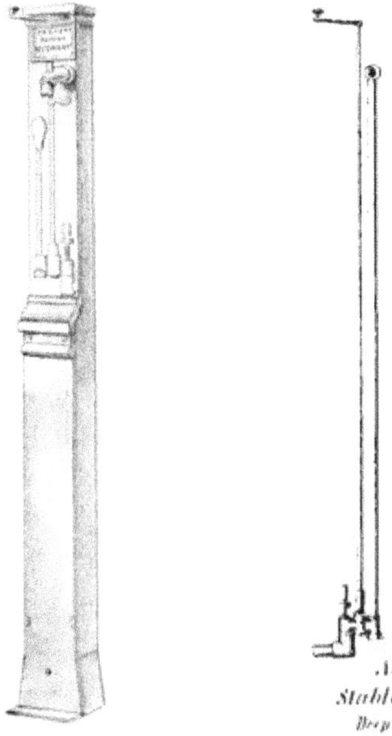

No. 277
Stop Cock Box (Ornamented Lid)

No. 278
Stop Cock Box
(Plain Lid)

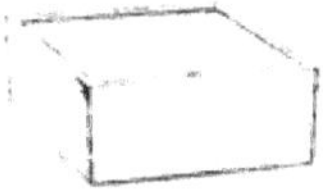

No. 279
Gas Stop Box

No. 281
Pavement Pipe

No. 280
Drain Grate

No. 283
Stable Trap
(Deep Pattern)

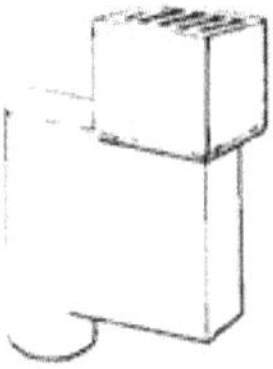

No. 282
Stable Trap
(Box Pattern)

No. 284
Stable Trap
(Shallow Pattern)

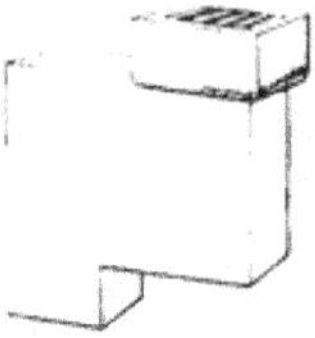

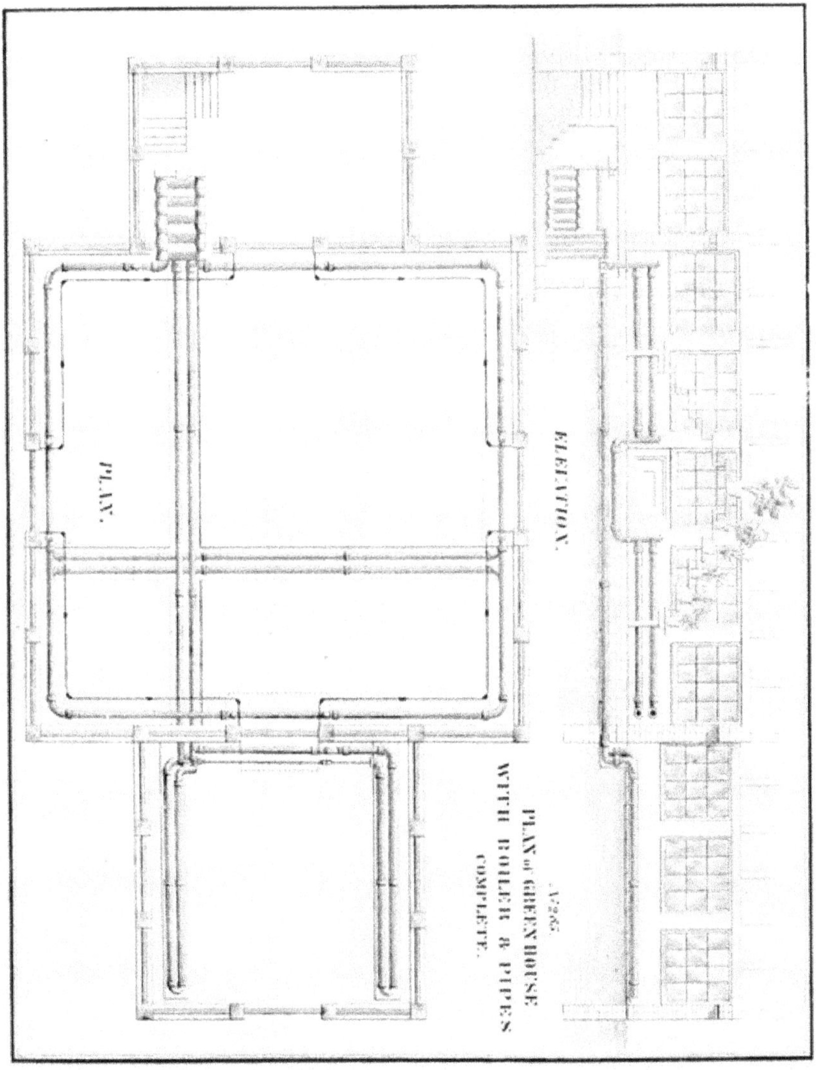

PLAN.
ELEVATION.
N° 25.
PLAN of GREEN HOUSE
WITH BOILER & PIPES
COMPLETE.

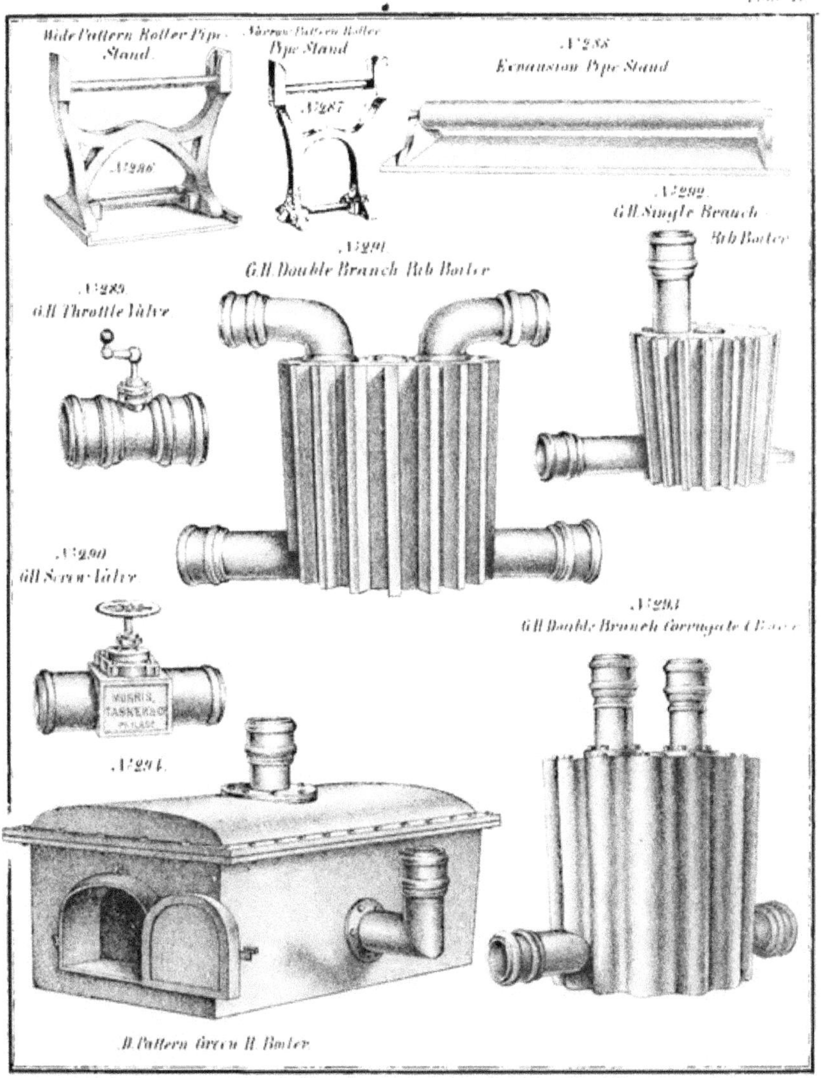
Wide Pattern Roller Pipe Stand.
No. 286
Pipe Stand
No. 287
No. 288
Expansion Pipe Stand
No. 292.
G.H. Single Branch
Rib Boiler
No. 291.
G.H. Double Branch Rib Boiler
No. 289.
G.H. Throttle Valve.
No. 290
G.H. Screw Valve
No. 293
G.H. Double Branch
MORRIS, TASKER & Co. PHILADA.
No. 294.
B. Pattern Green H. Boiler

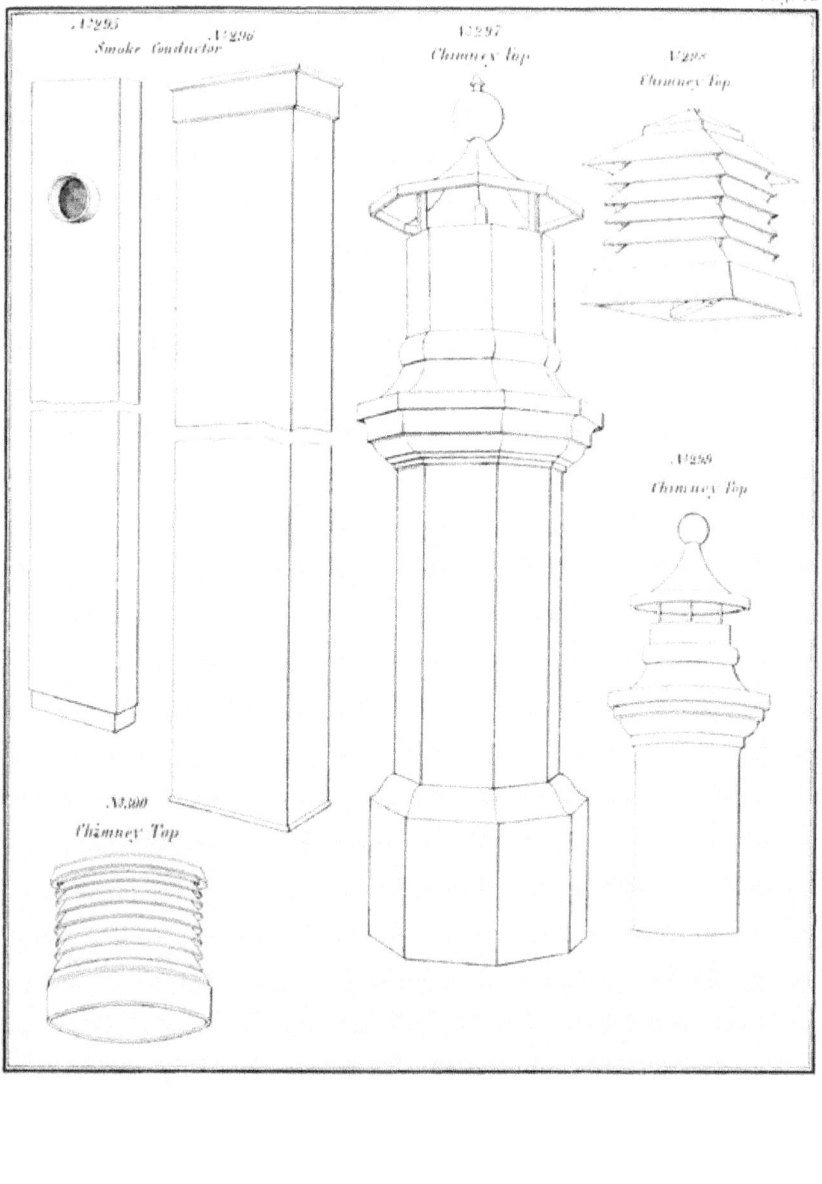

No. 295
Smoke Conductor
No. 296
No. 297
Chimney Top
No. 298
Chimney Top
No. 299
Chimney Top
No. 300
Chimney Top

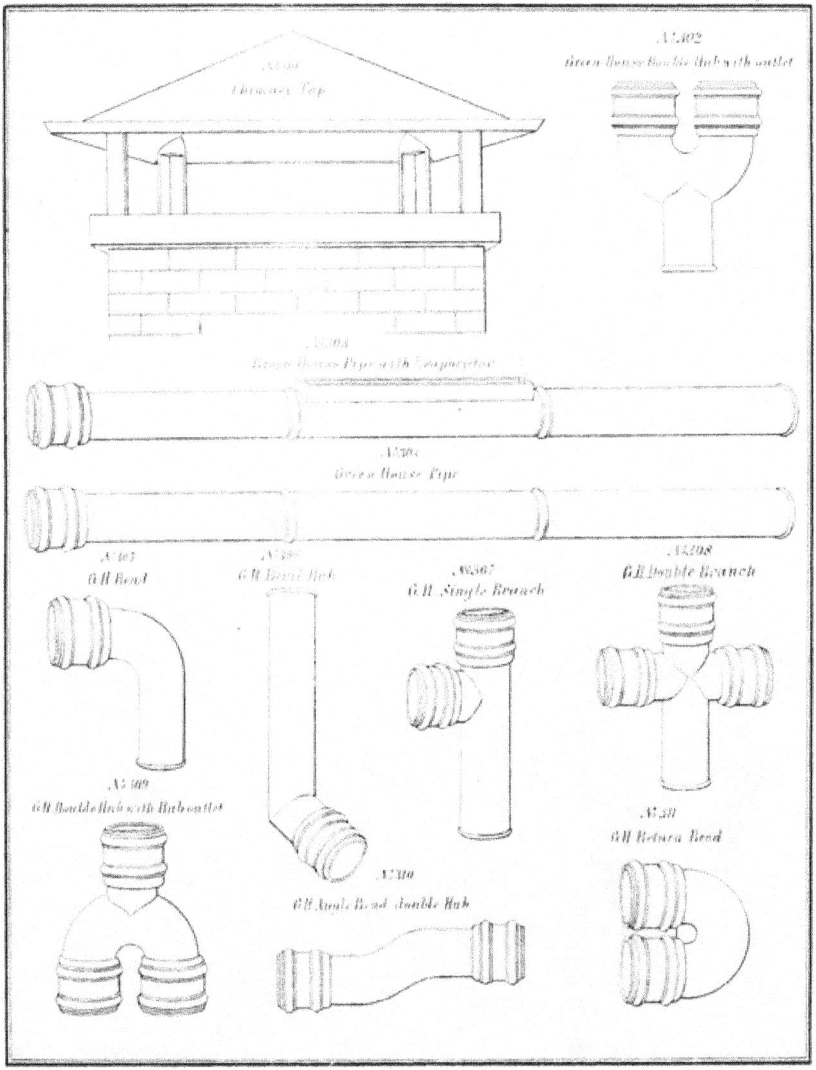
Chimney Top
N° 302
Green House Double Hub with outlet
Green House Pipe with Evaporator
Green House Pipe
G.H. Bend
G.H. Bend Hub
N° 307
G.H. Single Branch
N° 308
G.H. Double Branch
N° 309
G.H. Double Hub with Hub outlet
N° 310
G.H. Angle Bend double Hub
N° 311
G.H. Return Bend

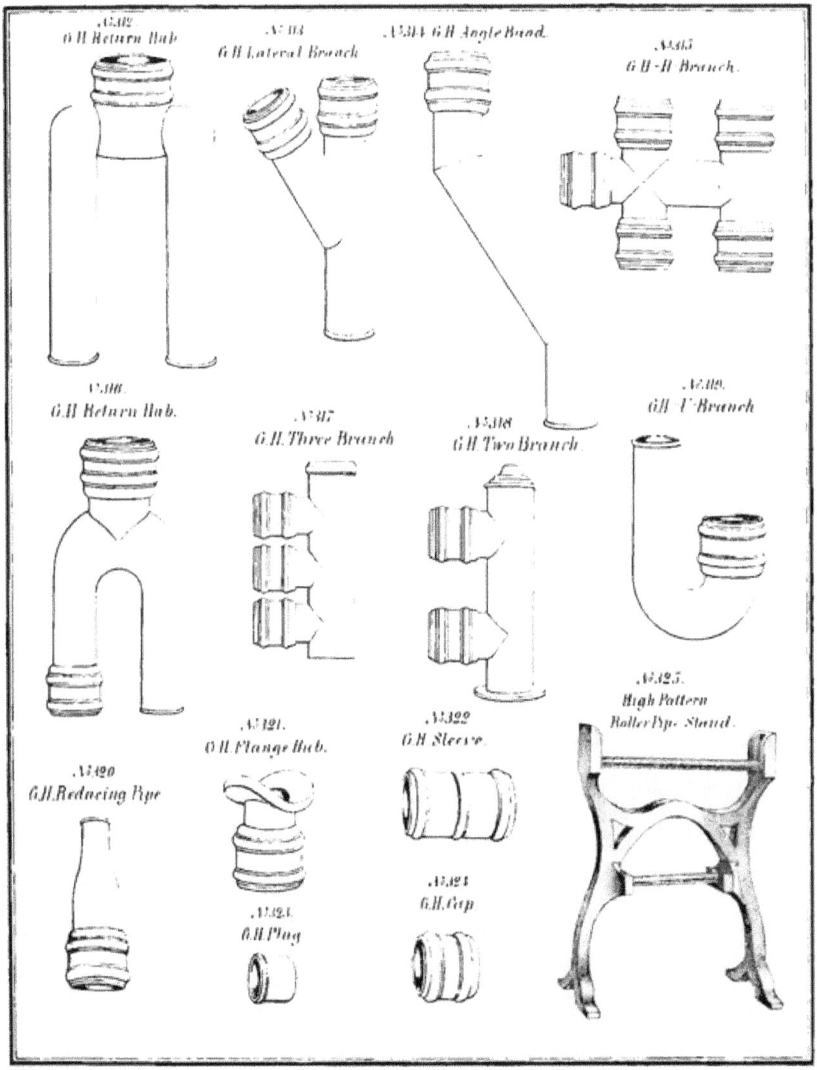
No. 312. G.H. Return Hub
No. 313. G.H. Lateral Branch
No. 314. G.H. Angle Band.
No. 315. G.H. H Branch.
No. 316. G.H. Return Hub.
No. 317. G.H. Three Branch
No. 318. G.H. Two Branch.
No. 319. G.H. V Branch
No. 320. G.H. Reducing Pipe
No. 321. G.H. Flange Hub.
No. 322. G.H. Sleeve.
No. 323. G.H. Plug
No. 324. G.H. Cap
No. 325. High Pattern Roller Pipe Stand.

N° 326 Fire Front with Sliding Doors

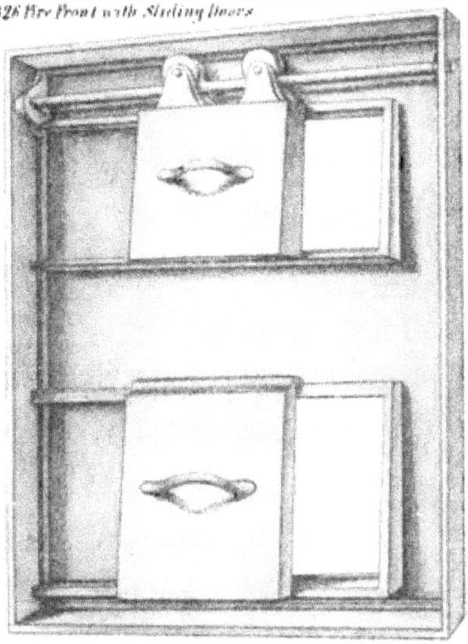

N° 327. Man Hole Door

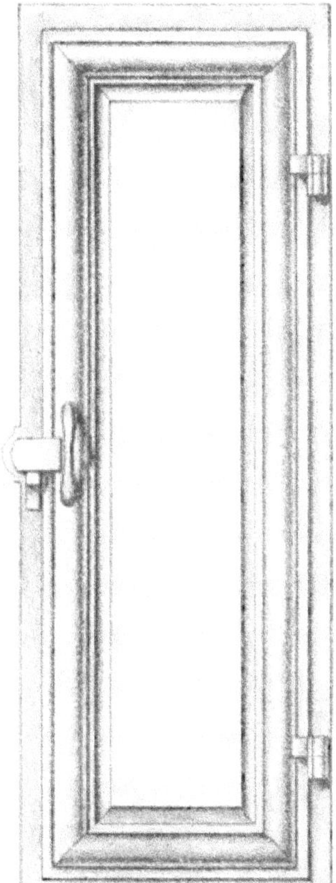

N° 328. Man Hole Door Small Pattern

N° 329 Flue Door

No. 330 Bake oven Door.

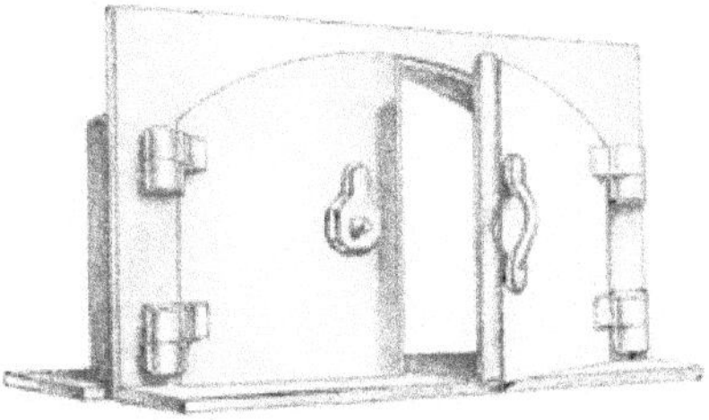

No. 331 Flue Door.

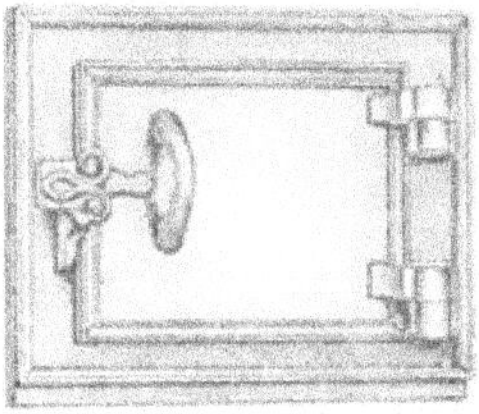

No. 332
Bake oven Door.

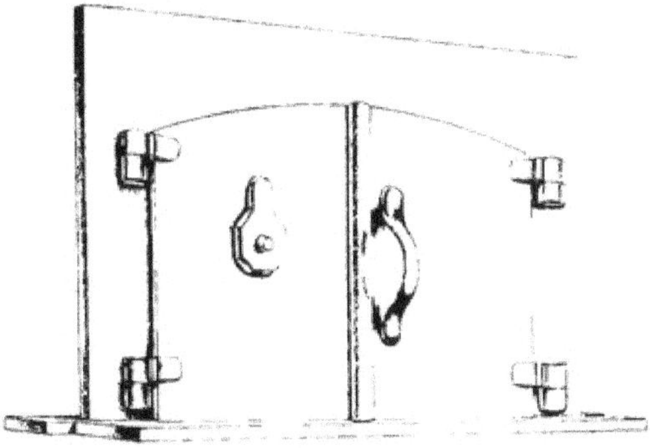

No. 333 Flue Door.

No. 334.
Bake oven Door.

No. 335
Fire Door.

No. 3436
Flue Door

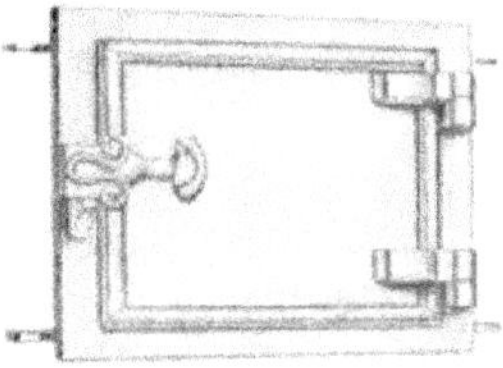

No. 3437
Flue Door

No. 3438
Flue Door

No. 3440
Oven Damper

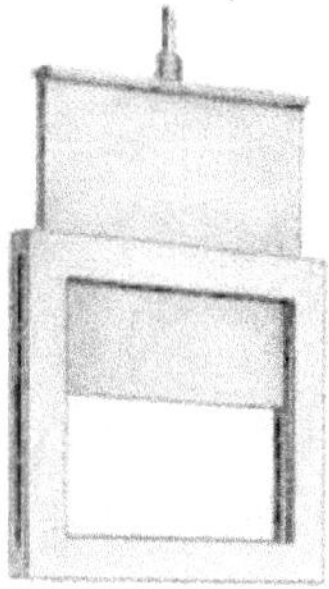

No. 3439
Flue Door

No. 3441
Damper

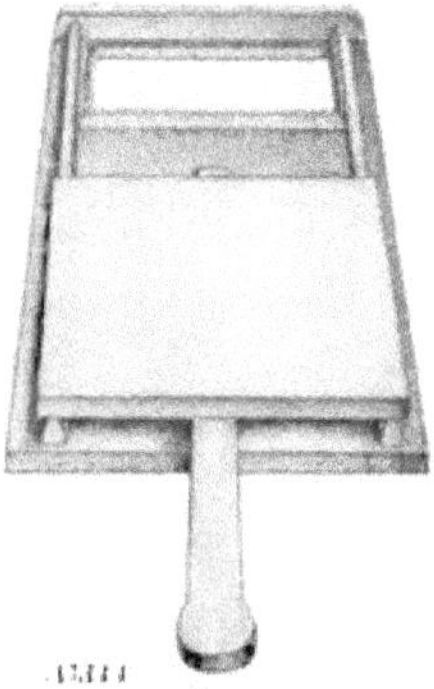

No. 3442
Damper

No. 3444
Iron Window Sash

No. 3443
Iron Window Sash

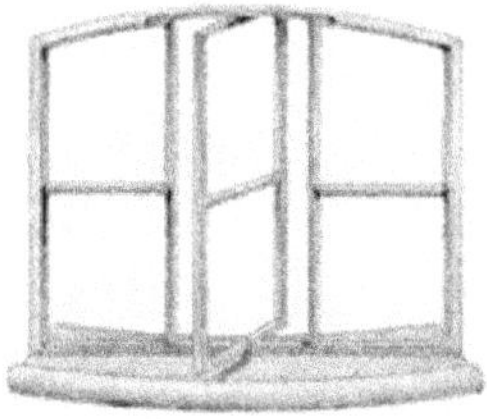

No. 345
Iron Window Sash

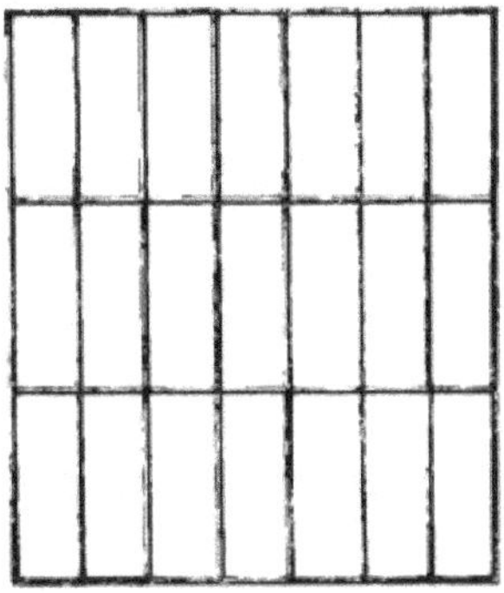

No. 346
Iron Window Sash

No. 353 Boot Scraper

No. 354
Scraper

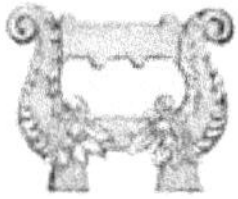

No. 348
Vault Ring

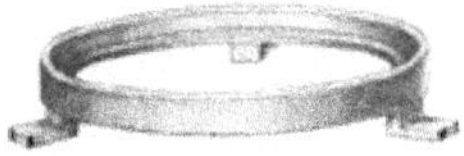

No. 355
Scraper

No. 347
Iron Window Sash

No. 349
Vault Cover

No. 356
Dumb Bell

No. 350
Vault Grate

No. 357
Door Roller

No. 351
Vault Ring with Cover

No. 352
Sewer Inlet

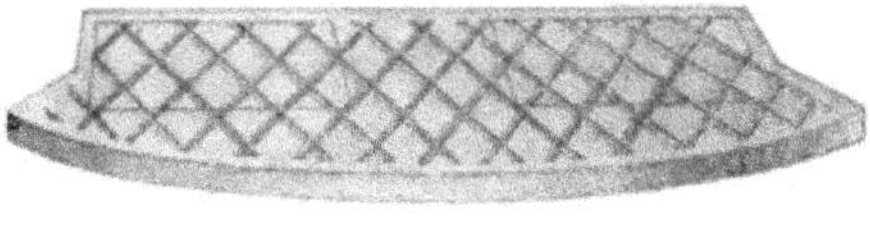

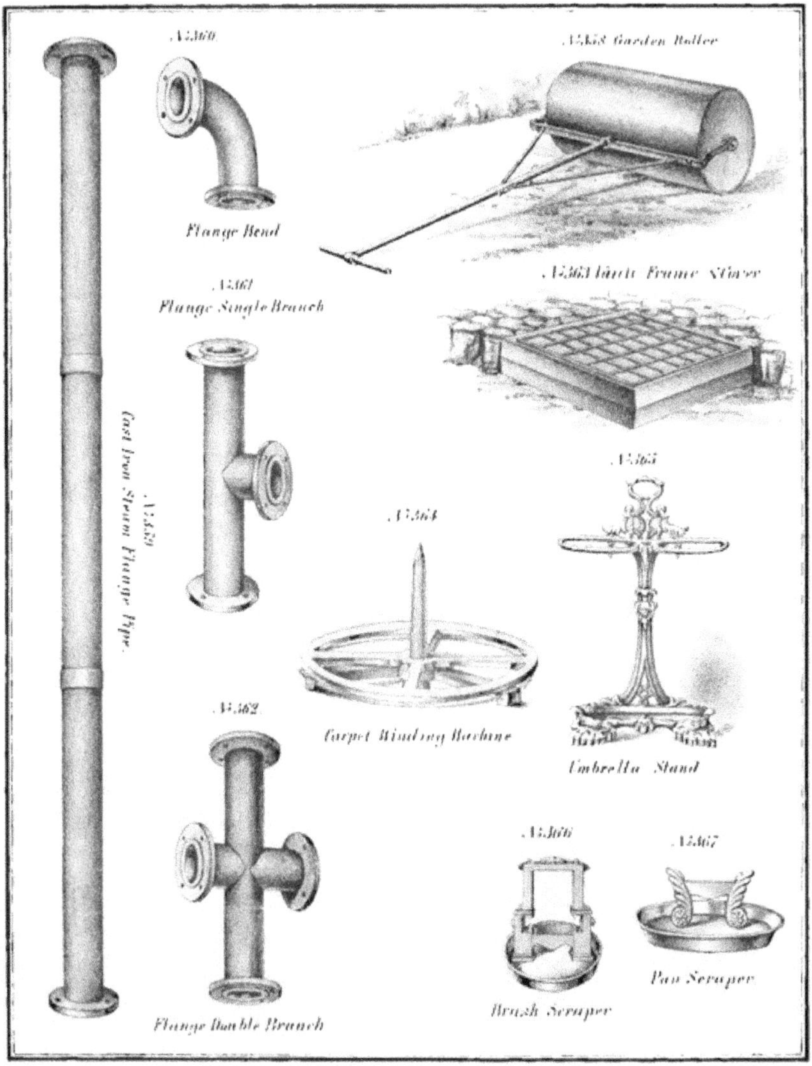
No. 359
Cast Iron Steam Flange Pipe.
No. 360.
Flange Bend
No. 361
Flange Single Branch
No. 362.
Flange Double Branch
No. 358 Garden Roller
No. 363 Earth Frame Sifter
No. 364
Carpet Winding Machine
No. 365
Umbrella Stand
No. 366
Brush Scraper
No. 367
Pan Scraper

No. 368 Garden Chair

No. 371

Garden Stool

No. 369 Garden Chair.

No. 372

Table Leg

No. 370

Double Garden Chair

No. 373

Umbrella Stand

No. 374

Mirror Frame

No. 376

Spittoon

No. 377

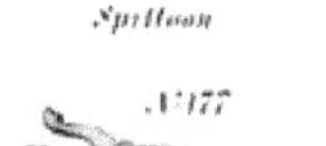

Boot Jack

No. 375

Mirror Frame

No. 478
Balcony Bracket

No. 479
Mantel Bracket

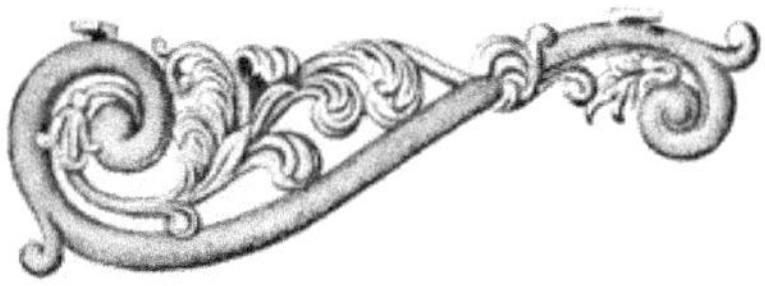

No. 480
Mantel Bracket

No. 481
Mantel Bracket

No. 482
Mantel Bracket

No. 483

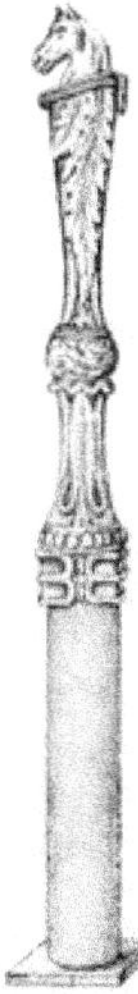

No. 484

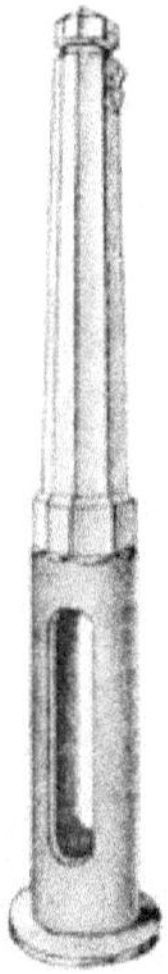

No. 485

No. 486

Hitching Posts

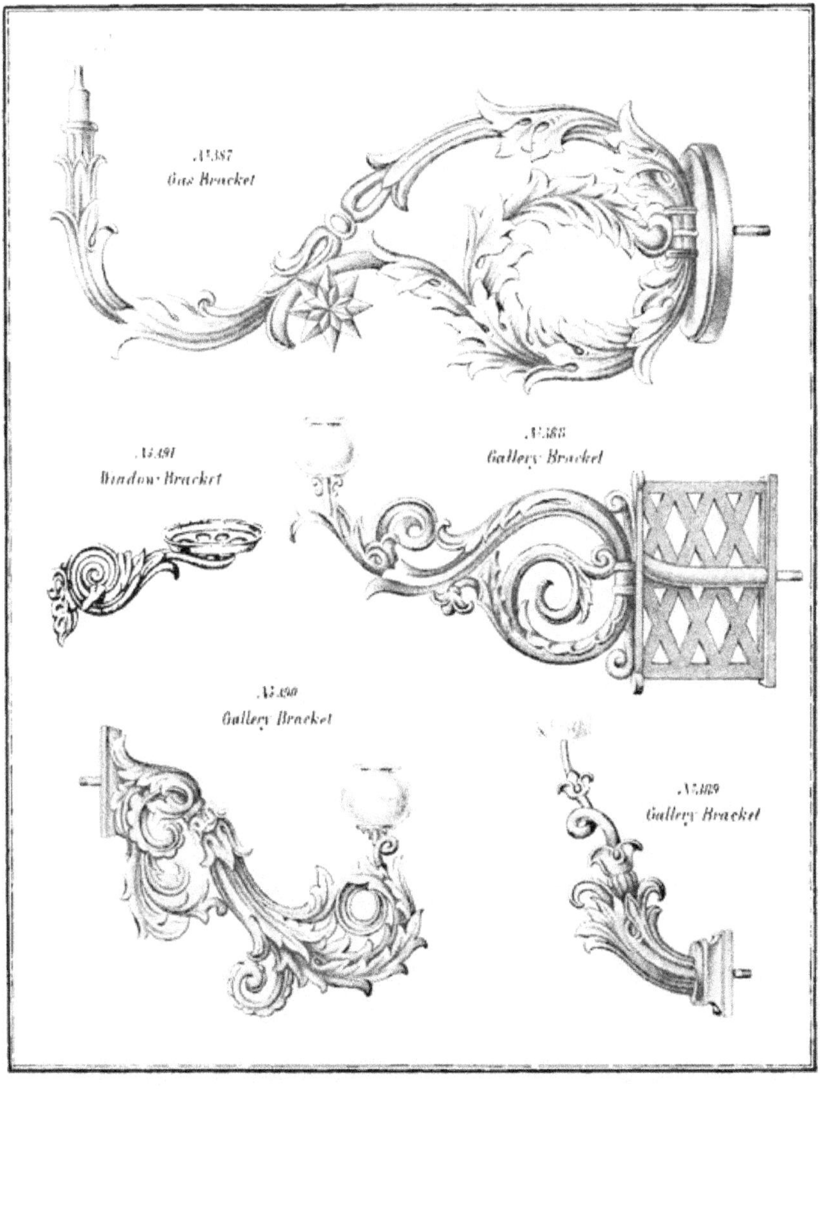
No. 387
Gas Bracket
No. 391
Window Bracket
No. 388
Gallery Bracket
No. 390
Gallery Bracket
No. 389
Gallery Bracket

N° 392
Capital Bracket
N° 393
Gallery Bracket.
N° 394
Mantel Bracket
N° 397
N° 395
Mantel Bracket
N° 396
Mantel Bracket
Pedestal Lamp

MORRIS, TASKER & Co.s ILLUSTRATED CATALOGUE.

Plate

Plate 54 a

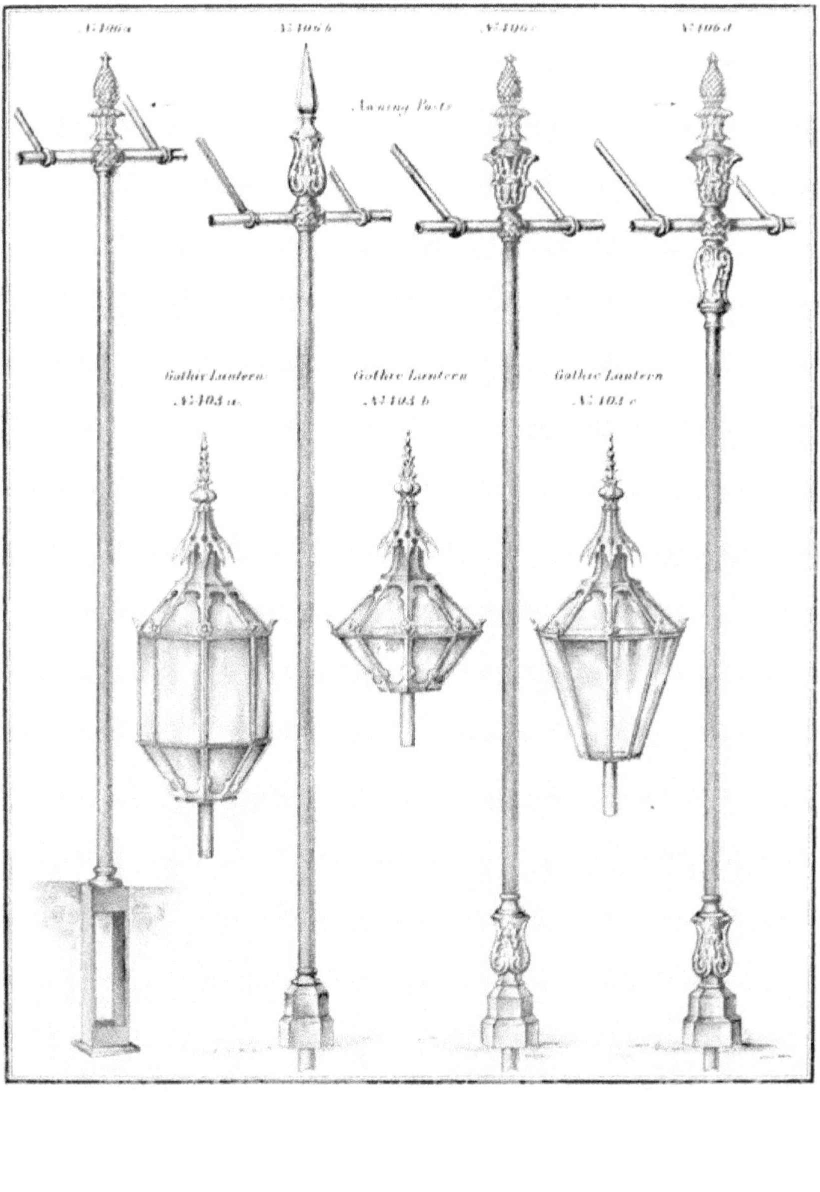
No. 106 a
No. 106 b
No. 106 c
No. 106 d
Awning Posts
Gothic Lantern
No. 103 a
Gothic Lantern
No. 103 b
Gothic Lantern
No. 103 c

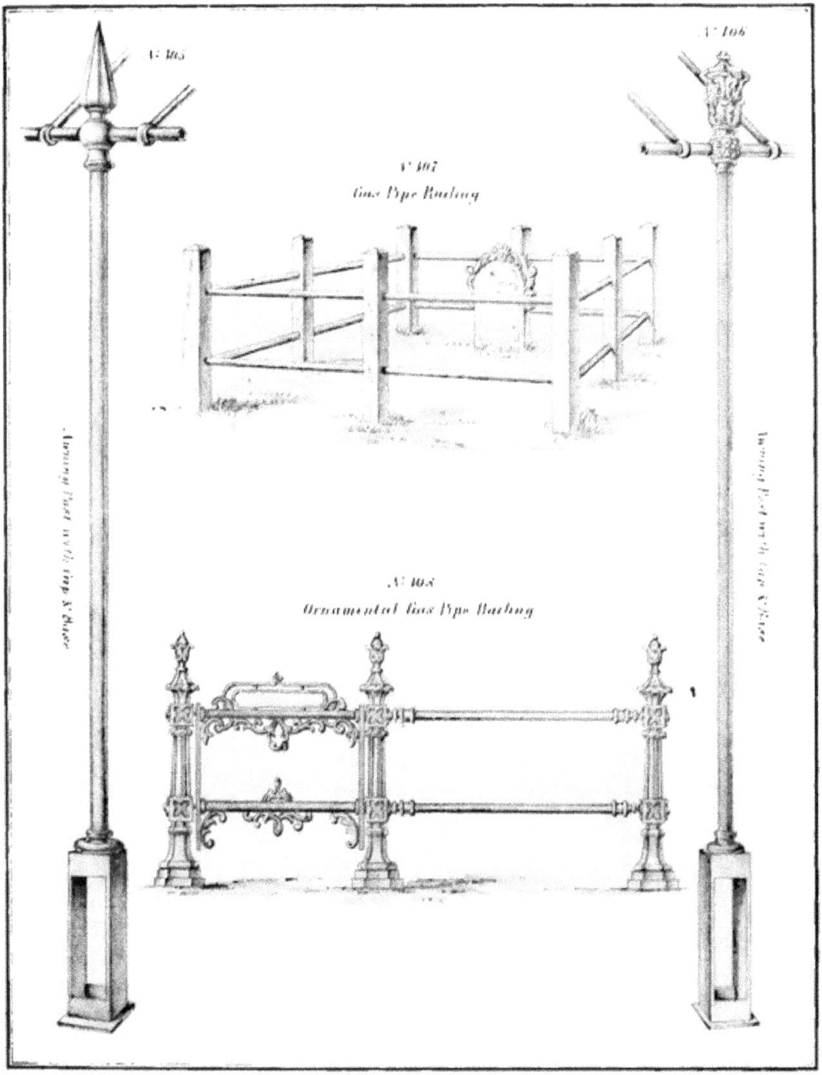

No. 105
Awning Post with Cap & Base
No. 106
Awning Post with Cap & Base
No. 107
Gas Pipe Railing
No. 108
Ornamental Gas Pipe Railing

Nº 409

Balcony Rail of Gas Pipe

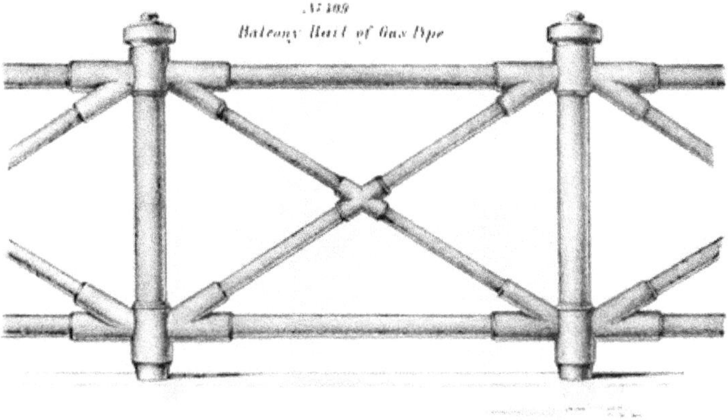

Nº 410

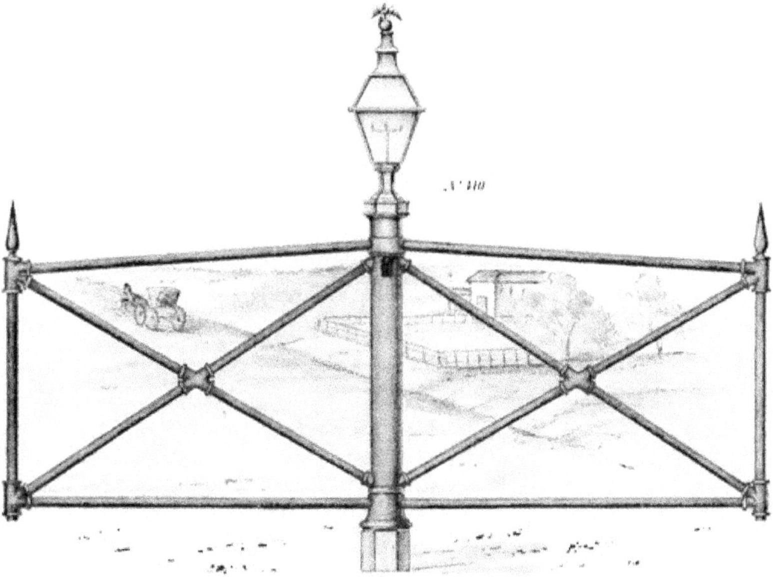

Rail Road & Turn Pike Gate of Gas Pipe

Plate 51a

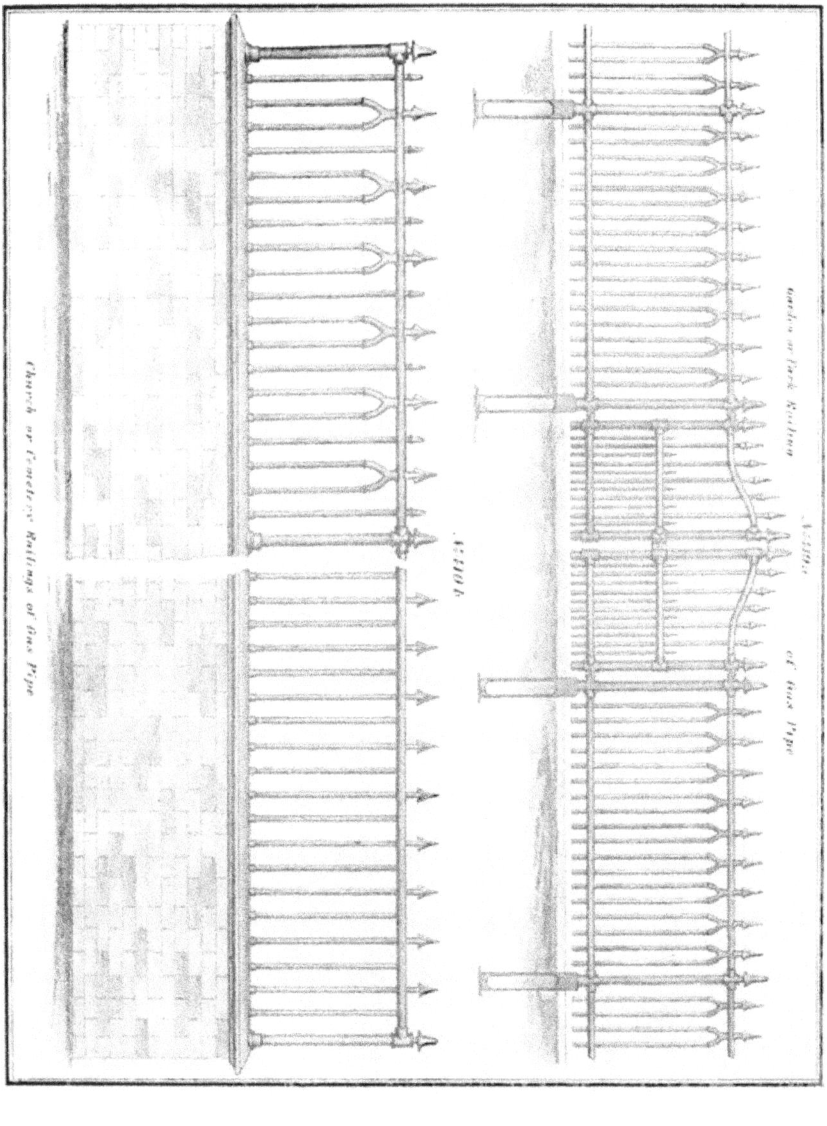
Church or Cemetery Railings of Gas Pipe
No 2104
Garden or Park Railing
No 2105
of Gas Pipe

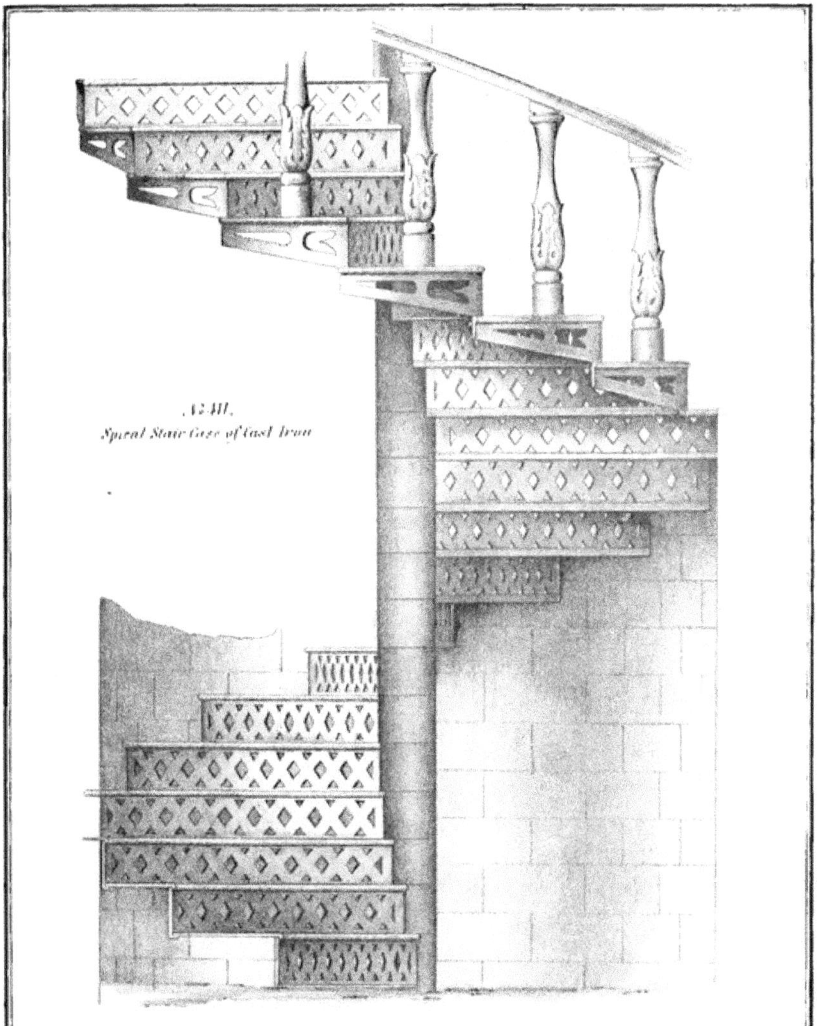

No. 411.

Spiral Stair Case of Cast Iron

No. 111 a

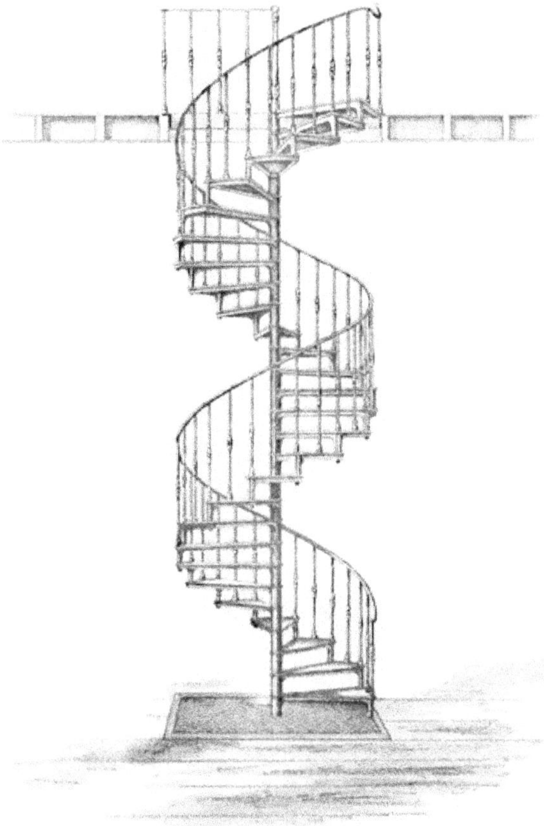

Spiral Stair Case or Fire Escape

Cast Iron Columns.

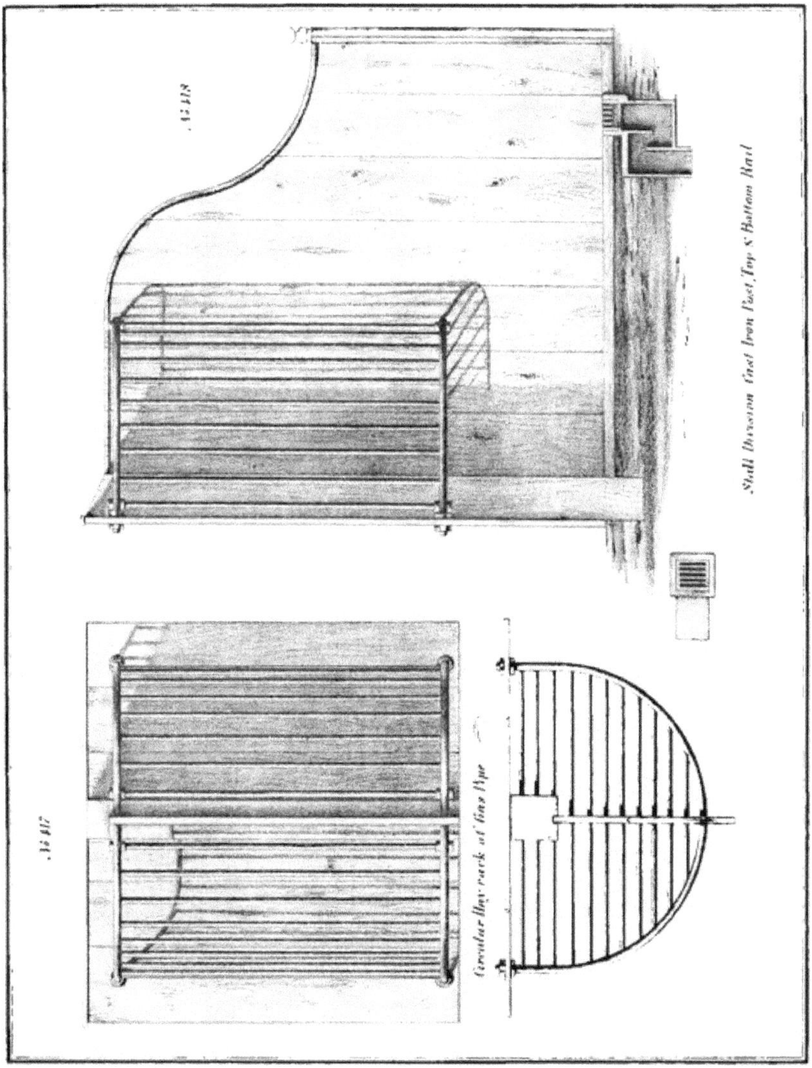
No 418
Stall Division Cast Iron Post, Top & Bottom Rail
No 417
Circular Hay-rack of Gas Pipe

No. 110.

Stable Arrangements of Iron Tube.

Front Elevation

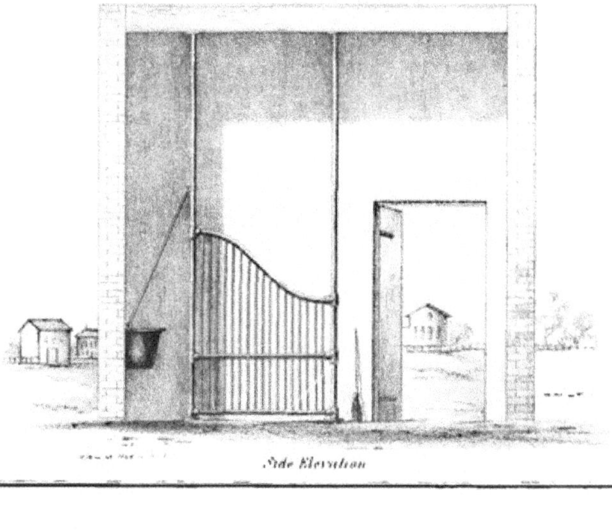

Side Elevation

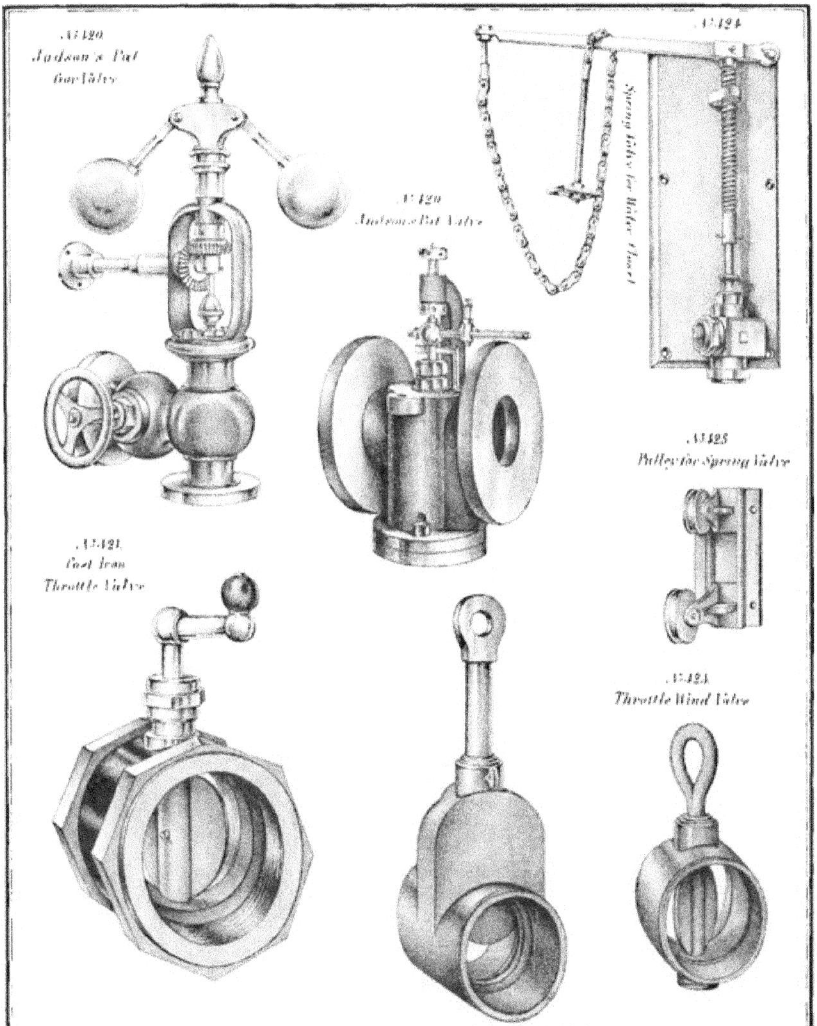
No. 420
Judson's Pat
Gov Valve
No. 420
Judson's Pat Valve
No. 424
Spring Valve for Water Closet
No. 425
Pulley for Spring Valve
No. 421
Cast Iron
Throttle Valve
No. 423
Throttle Wind Valve
No. 422 Slide Wind Valve

No 426

Tasker's Pat. Self Regulating Hot Water Furnace
Designed for Heating Public Buildings
& Private Residences

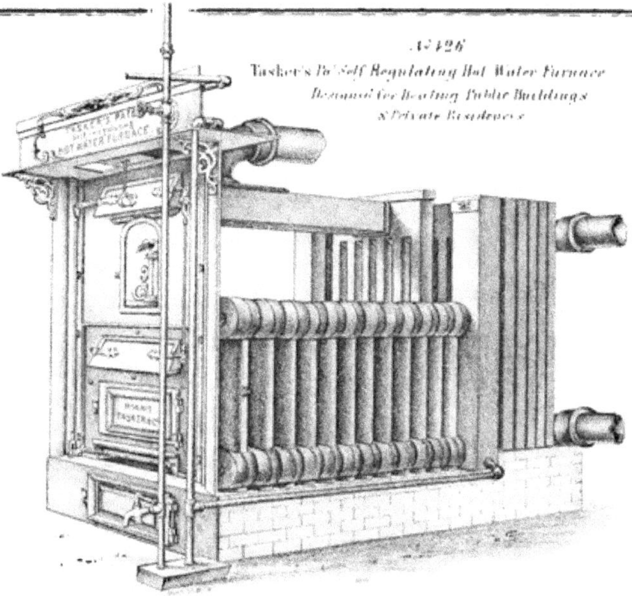

No 426
Side Section.

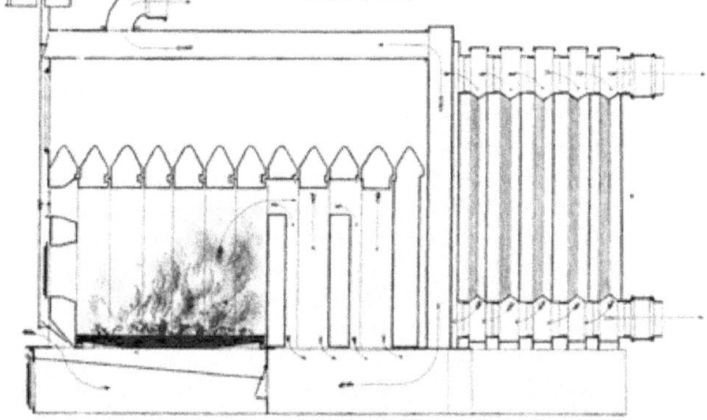

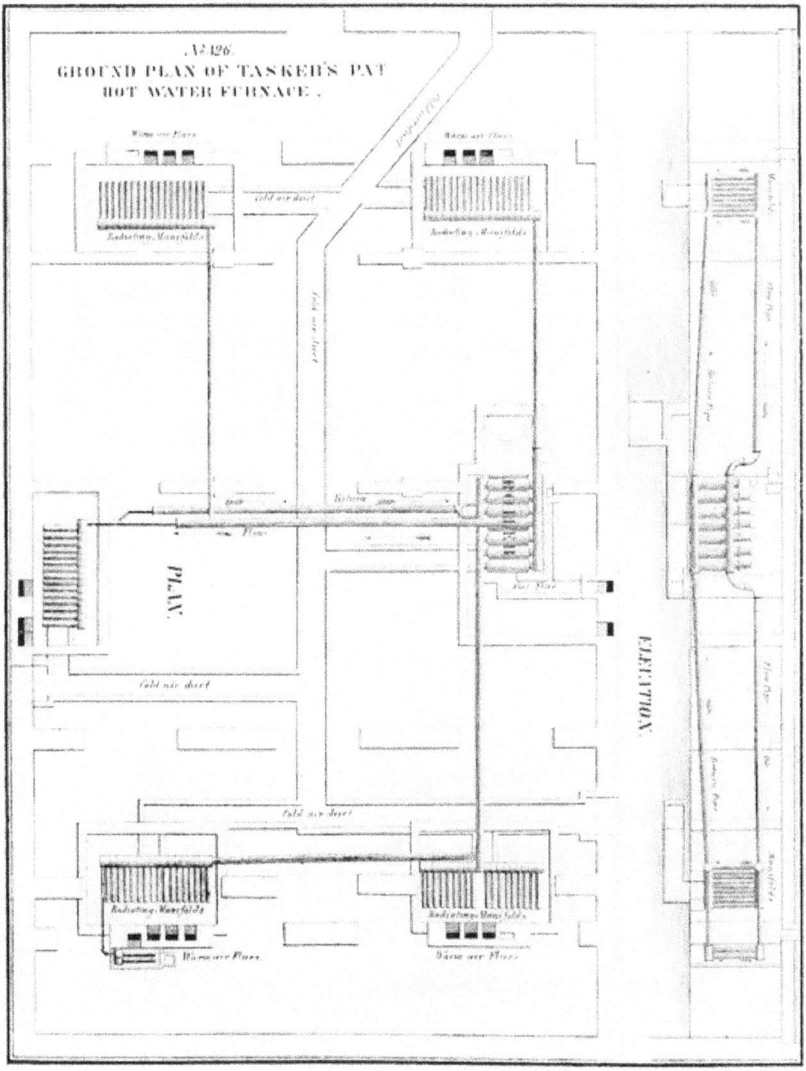
No. 126.
GROUND PLAN OF TASKER'S PAT
HOT WATER FURNACE.
Warm air Flues
Cold air duct
Radiating Manifolds
Return
Flow
PLAN.
Cold air duct
Warm air Flues
ELEVATION.

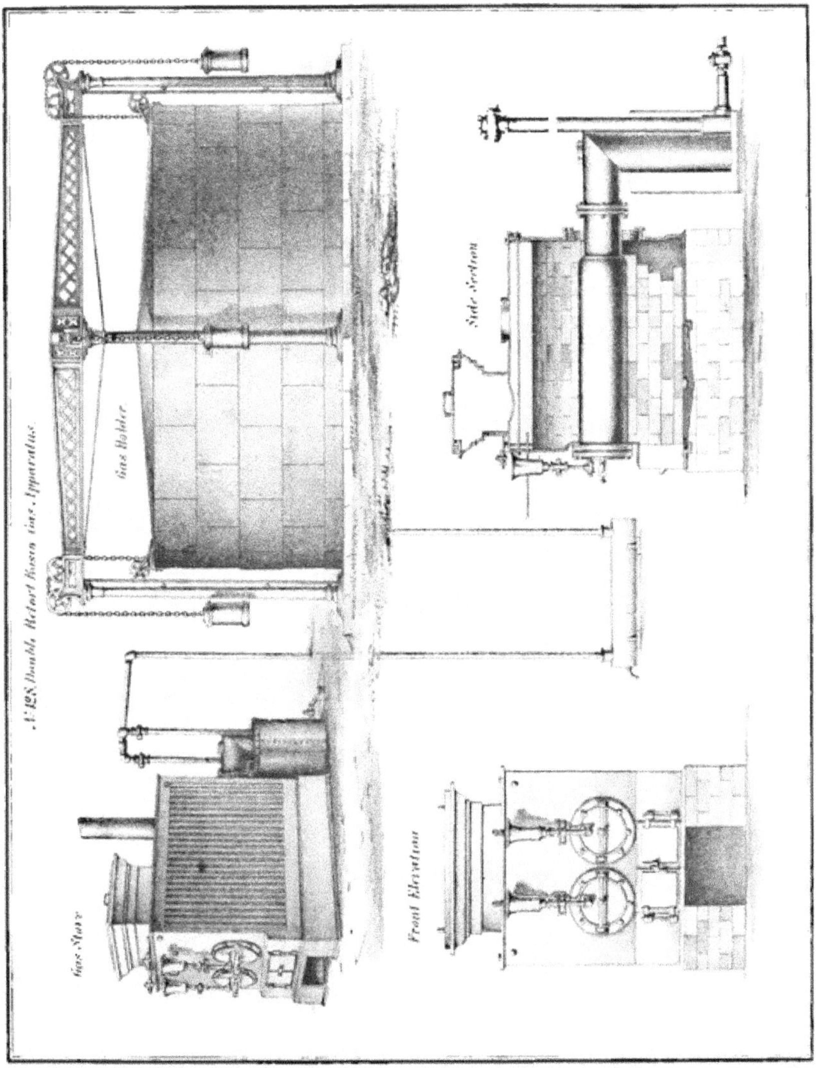

No. 128, Double Retort Rosin Gas Apparatus.

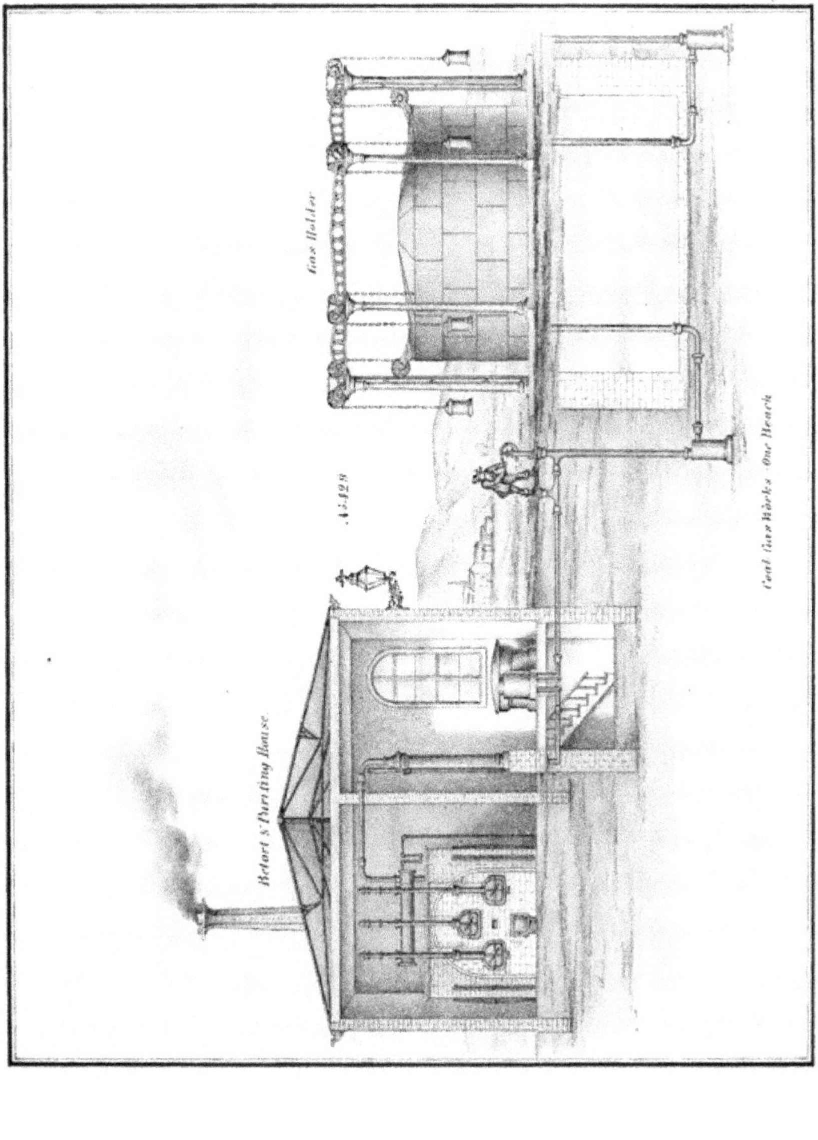
Retort & Purifying House.
Gas Holder
No 428
Coal Gas Works - One Bench

Telescope Gas Holder

No. 430

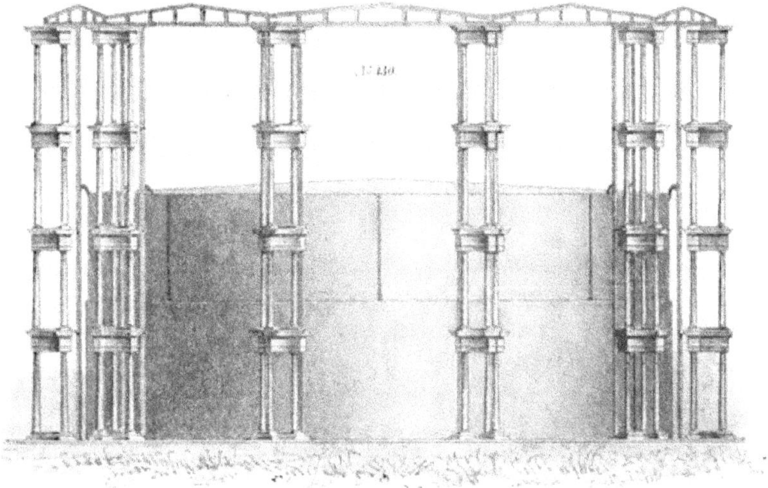

No. 432

Harbor Light

[illegible]

No. 431

Mast Gas Holder

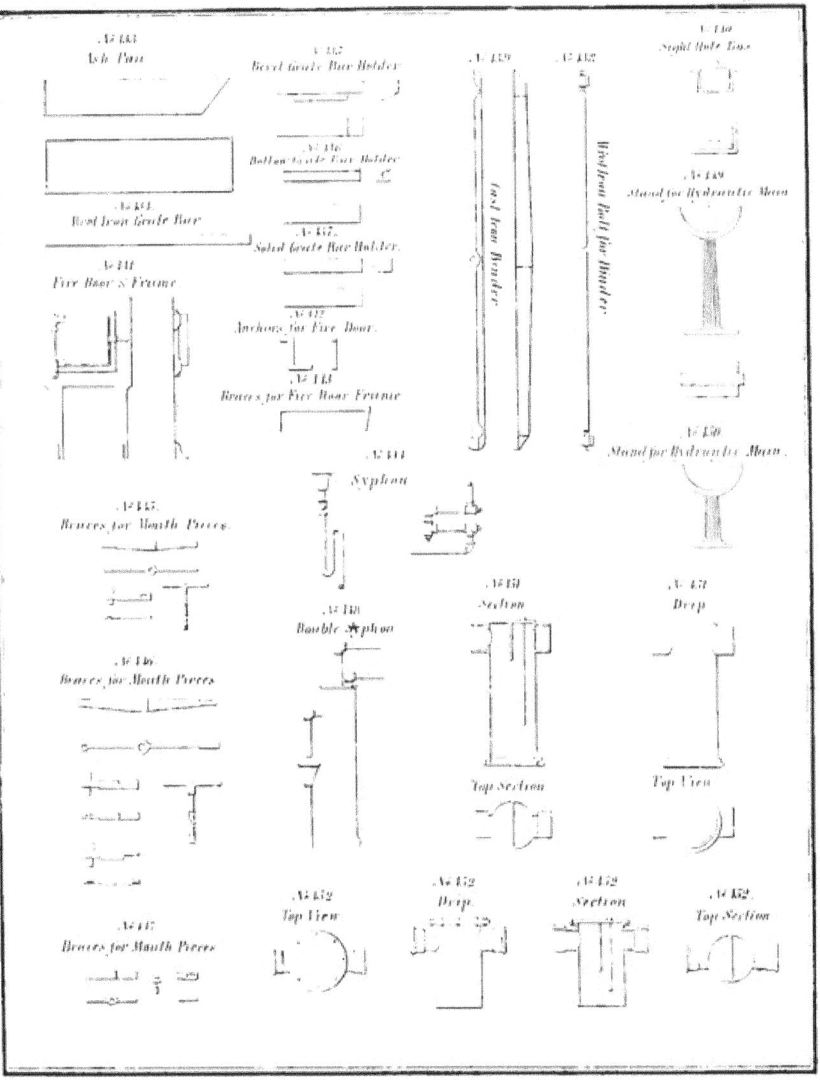
Ash Pan
Bevel Grate Bar Holder
Bottom Grate Bar Holder
Bent Iron Grate Bar
Solid Grate Bar Holder
Fire Door & Frame
Anchors for Fire Door
Braces for Fire Door Frame
Cast Iron Binder
Wrought Iron Bolt for Binder
Sight Hole Box
Stand for Hydraulic Main
Stand for Hydraulic Main
Syphon
Braces for Mouth Pieces
Braces for Mouth Pieces
Double Syphon
Section
Top Section
Drip
Top View
Braces for Mouth Pieces
Top View
Drip
Section
Top Section

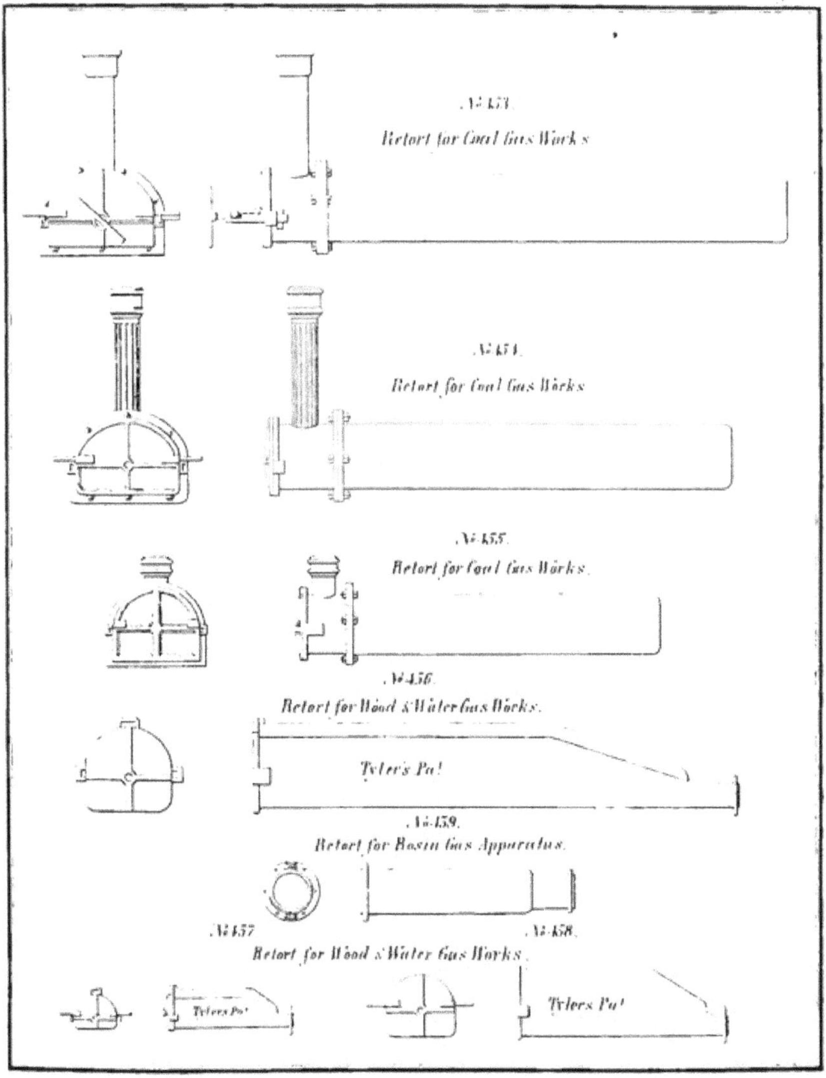
№ 453.
Retort for Coal Gas Works
№ 454.
Retort for Coal Gas Works
№ 455.
Retort for Coal Gas Works.
№ 456.
Retort for Wood & Water Gas Works.
Tyler's Pat
№ 459.
Retort for Rosin Gas Apparatus.
№ 457
№ 458.
Retort for Wood & Water Gas Works.
Tylers Pat
Tylers Pat

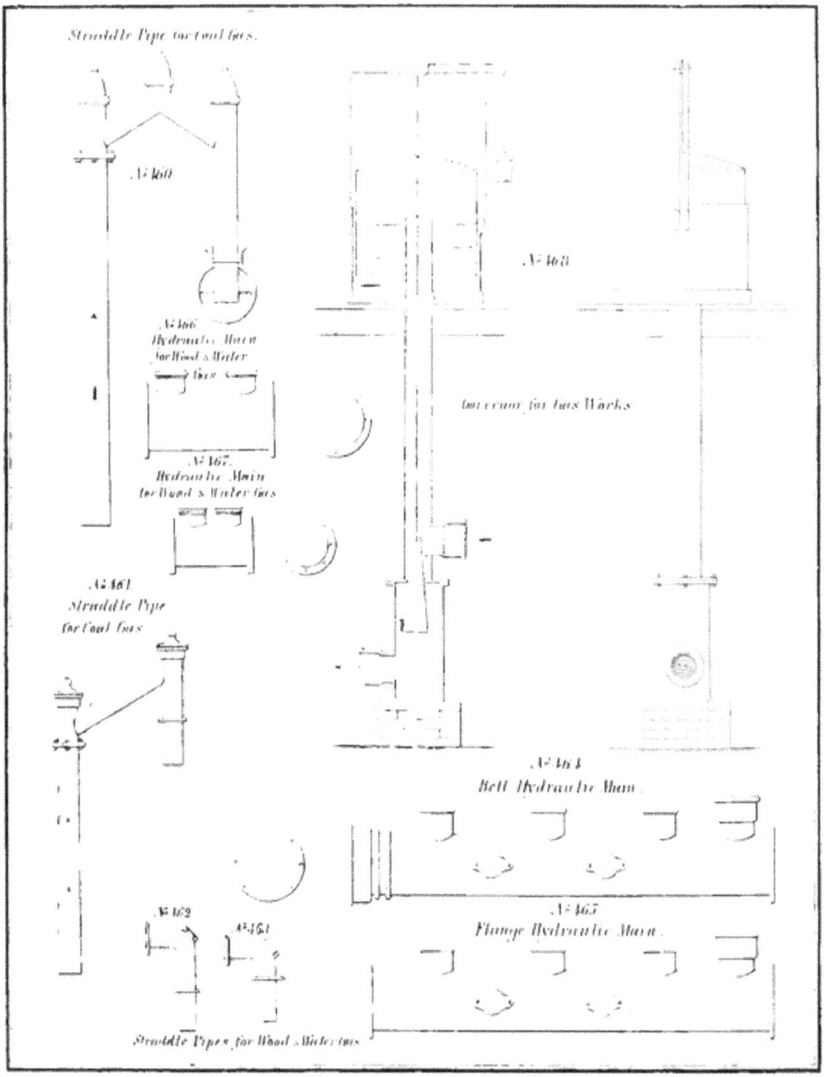
Straddle Pipe for Coal Gas.
No 460
No 466
Hydraulic Main
for Wood & Water
Gas
No 467.
Hydraulic Main
for Wood & Water Gas
No 461
Straddle Pipe
for Coal Gas
No 463
Governor for Gas Works
No 464
Bell Hydraulic Main.
No 465
Flange Hydraulic Main.
No 462
Straddle Pipes for Wood & Water Gas

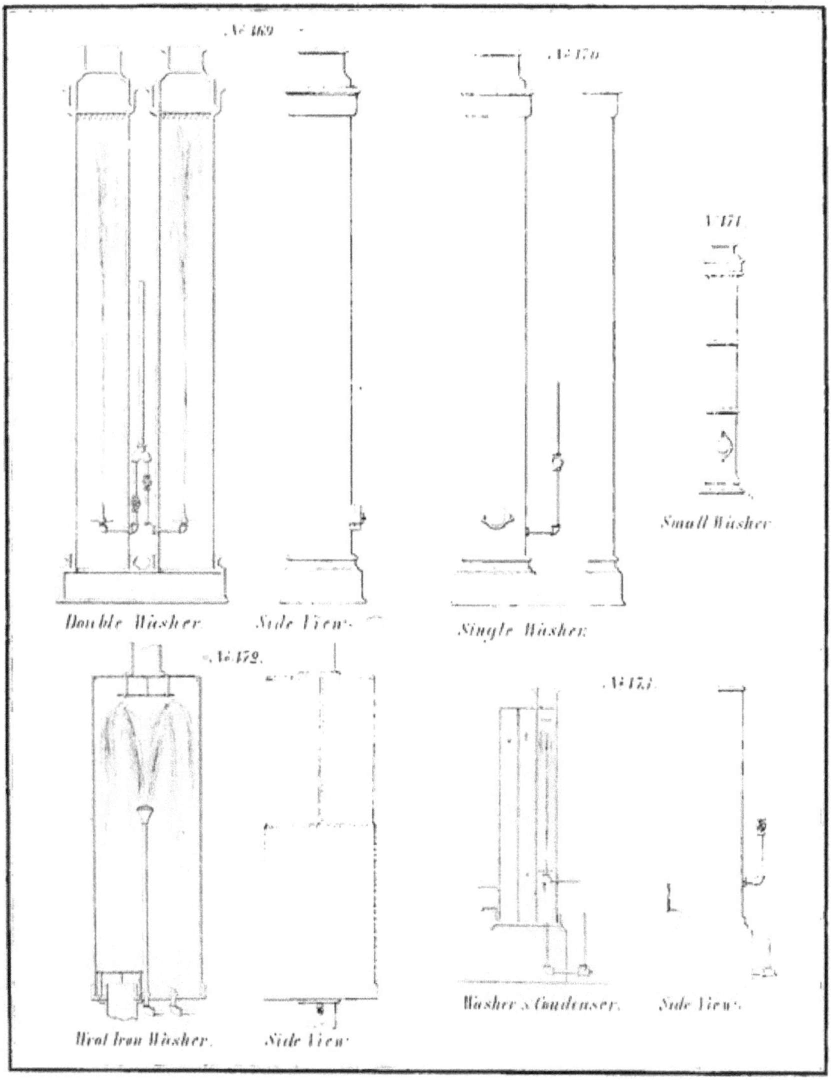
No. 469.
Double Washer.
Side View.
No. 470.
Single Washer.
No. 471.
Small Washer.
No. 472.
Wrot Iron Washer.
Side View.
No. 473.
Washer & Condenser.
Side View.

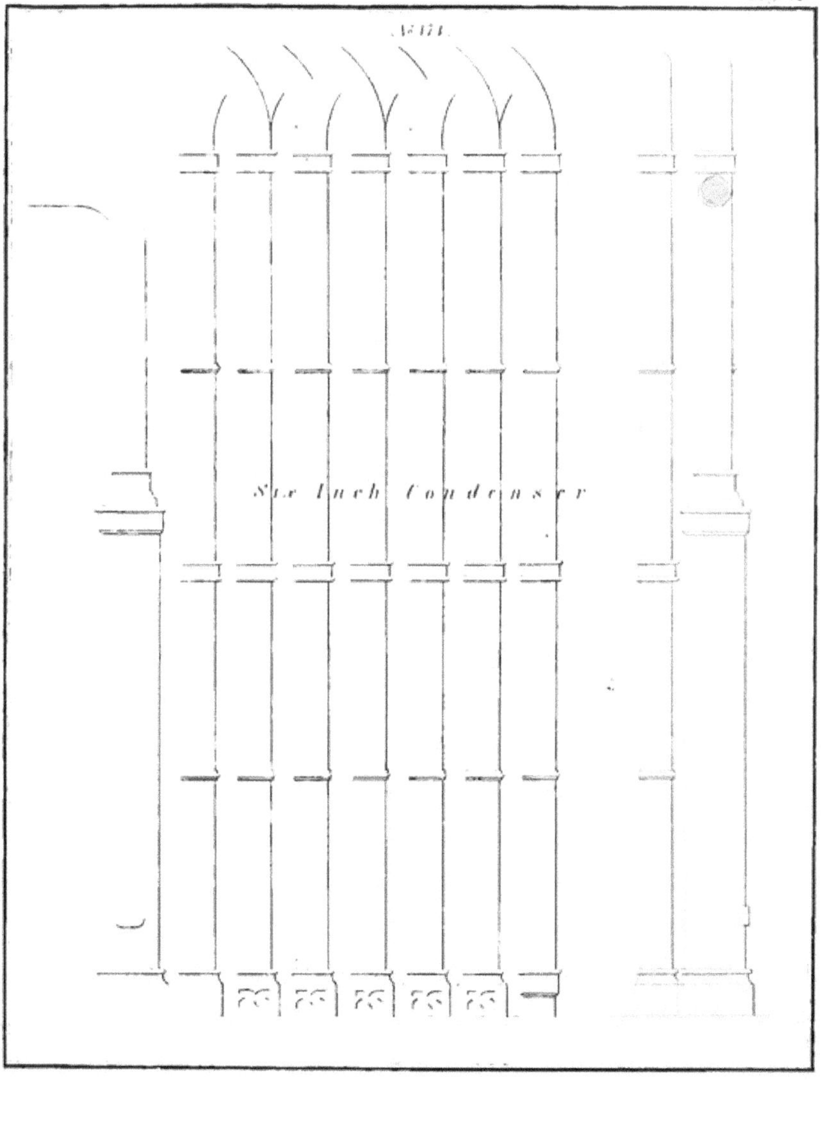
No. 171.
Six Inch Condenser

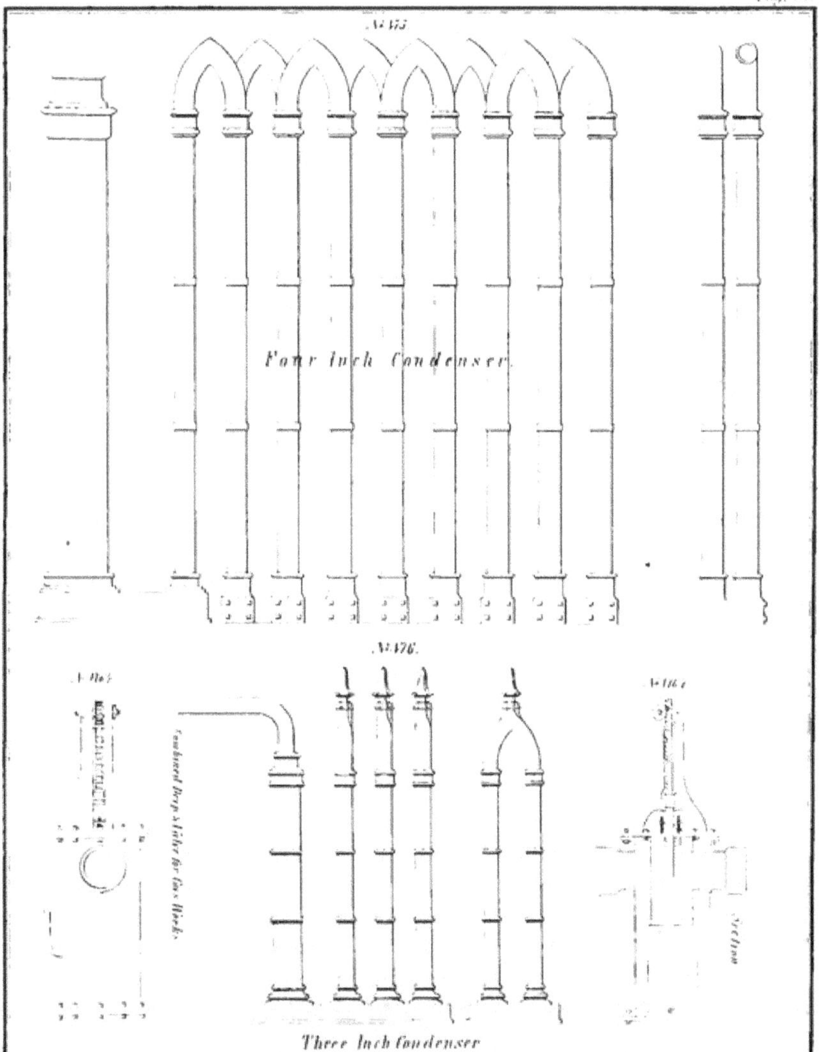
No. 475.
Four Inch Condenser.
No. 476.
Section
Three Inch Condenser

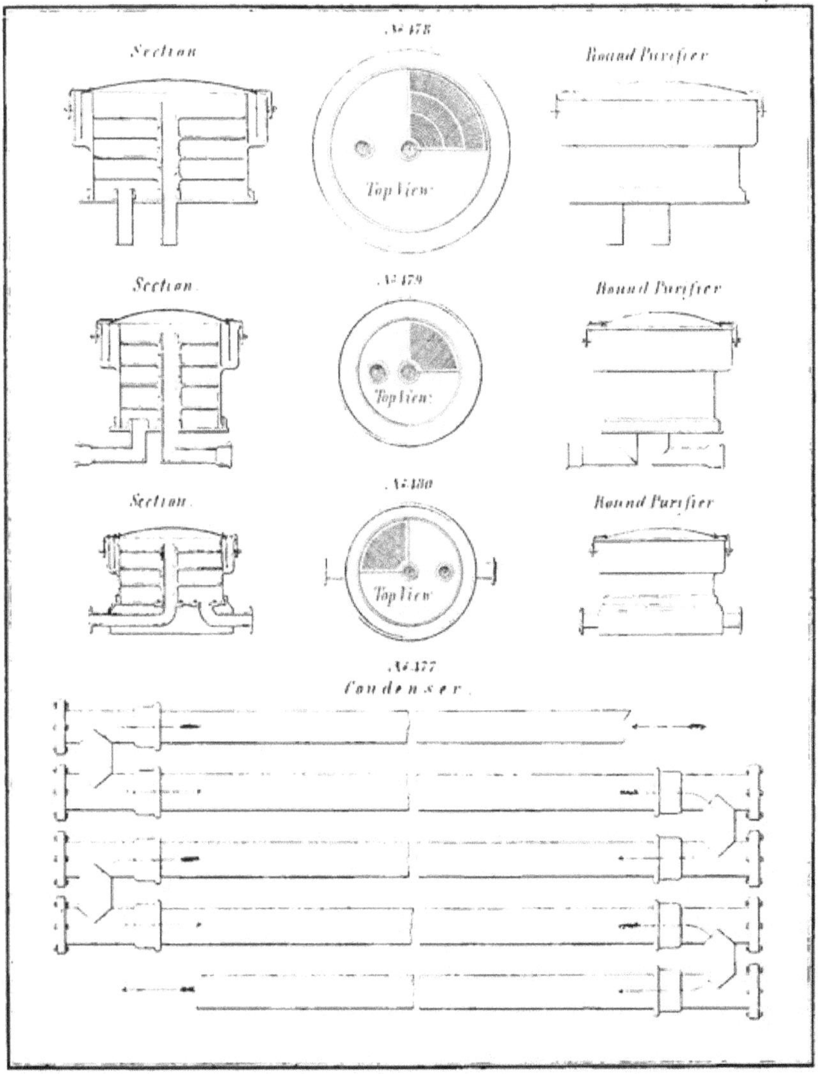
Section
No. 478
Round Purifier
Top View
Section.
No. 479
Round Purifier
Top View
Section.
No. 480
Round Purifier
Top View
No. 477
Condenser.

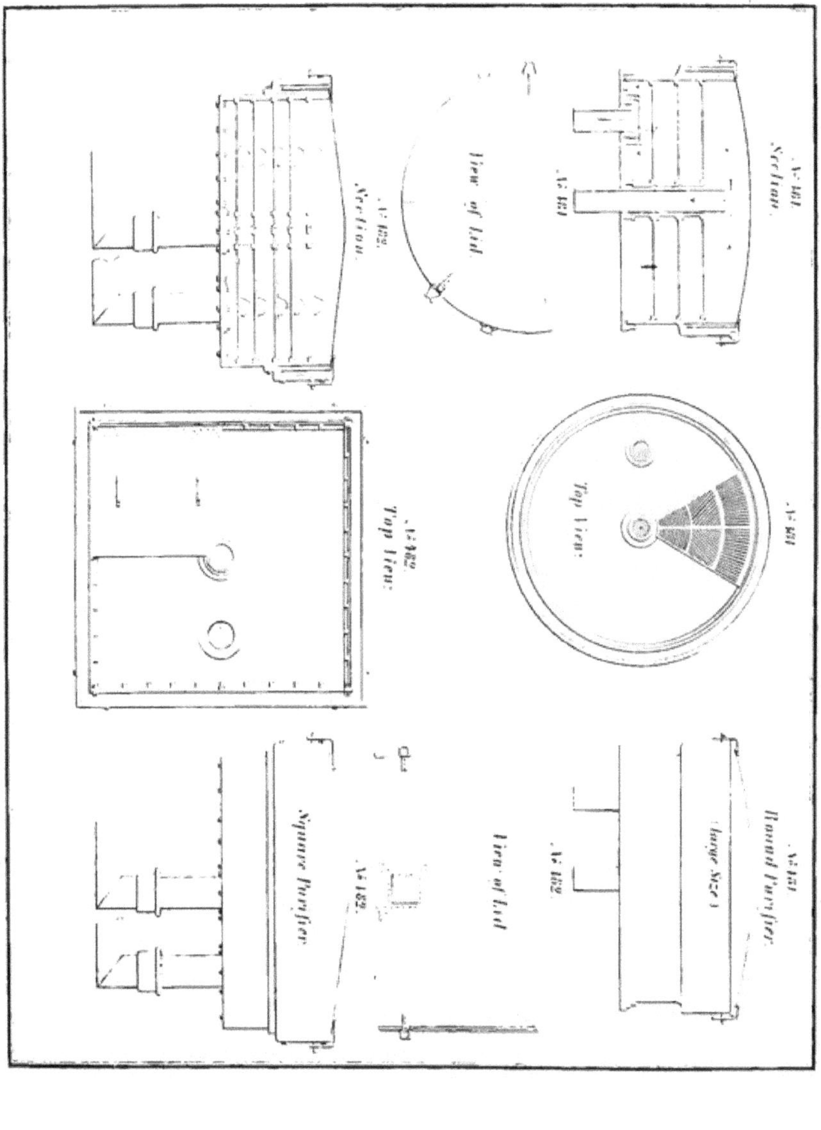
No 181
Section.
No 181
View of Lid
No 182
Section.
No 181
Top View.
No 182
Top View.
No 181
Round Purifier.
(Large Size)
No 182
View of Lid
No 182
Square Purifier

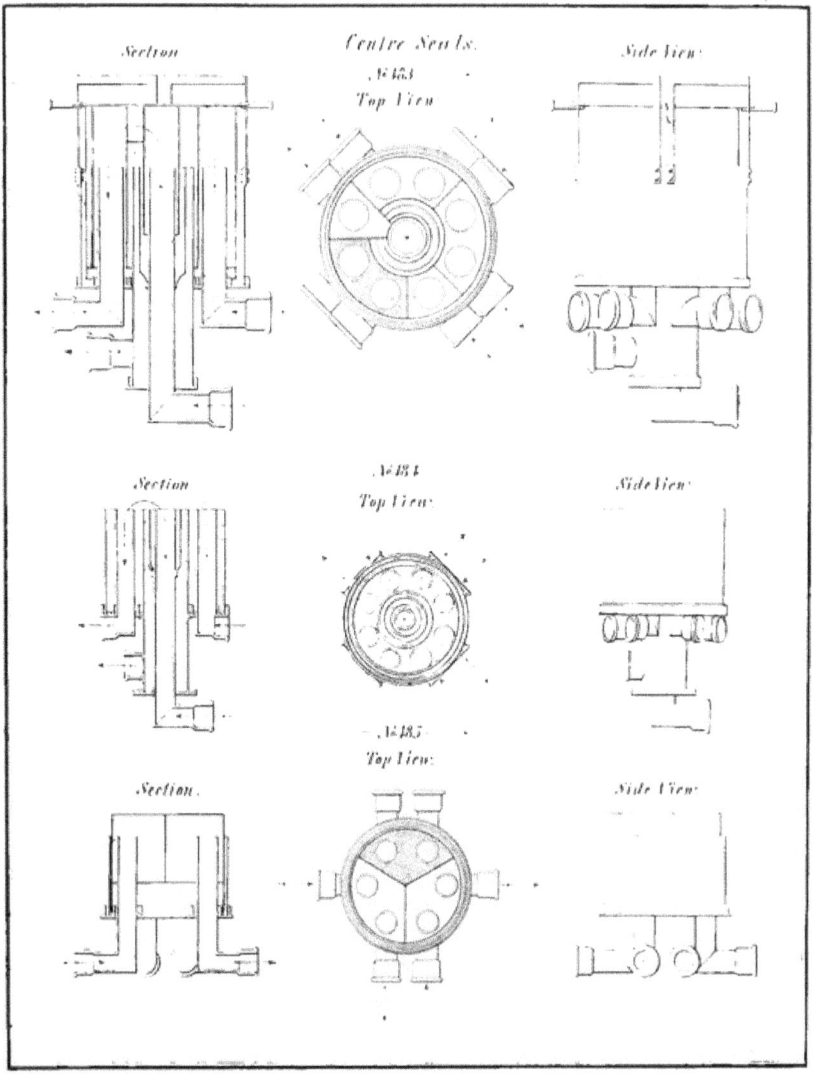
Centre Seals.
Section
No 483
Top View
Side View
Section
No 484
Top View
Side View
No 485
Top View
Section.
Side View

MALLEABLE IRON GAS FITTINGS.

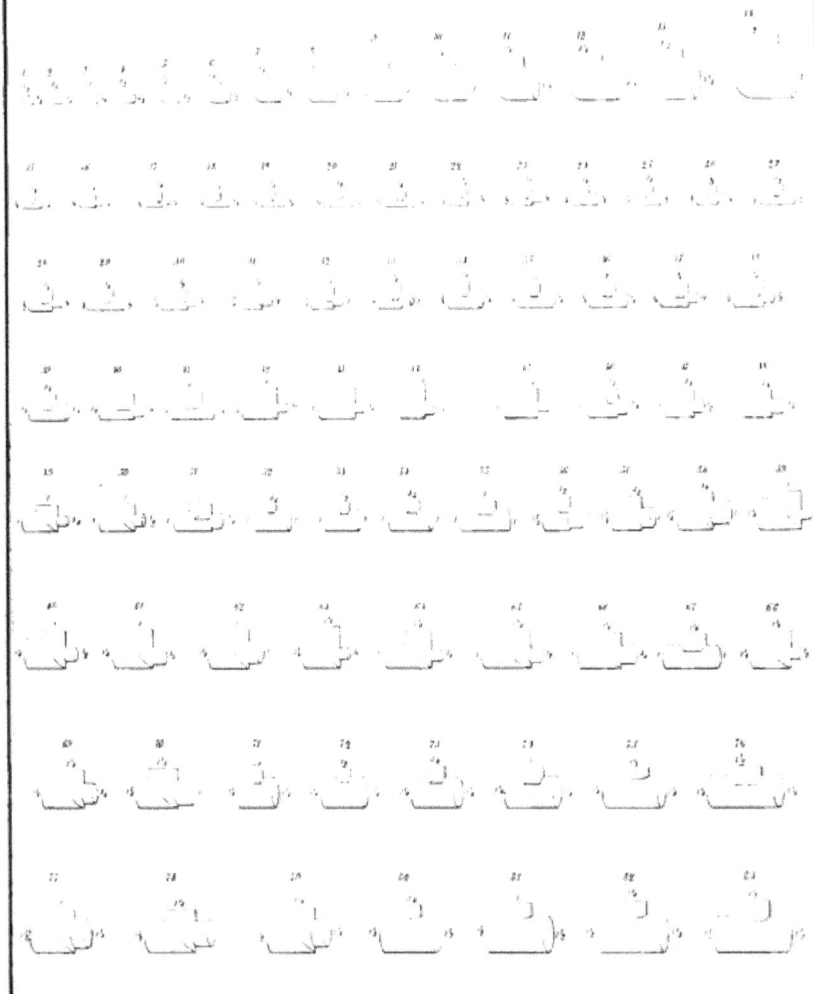

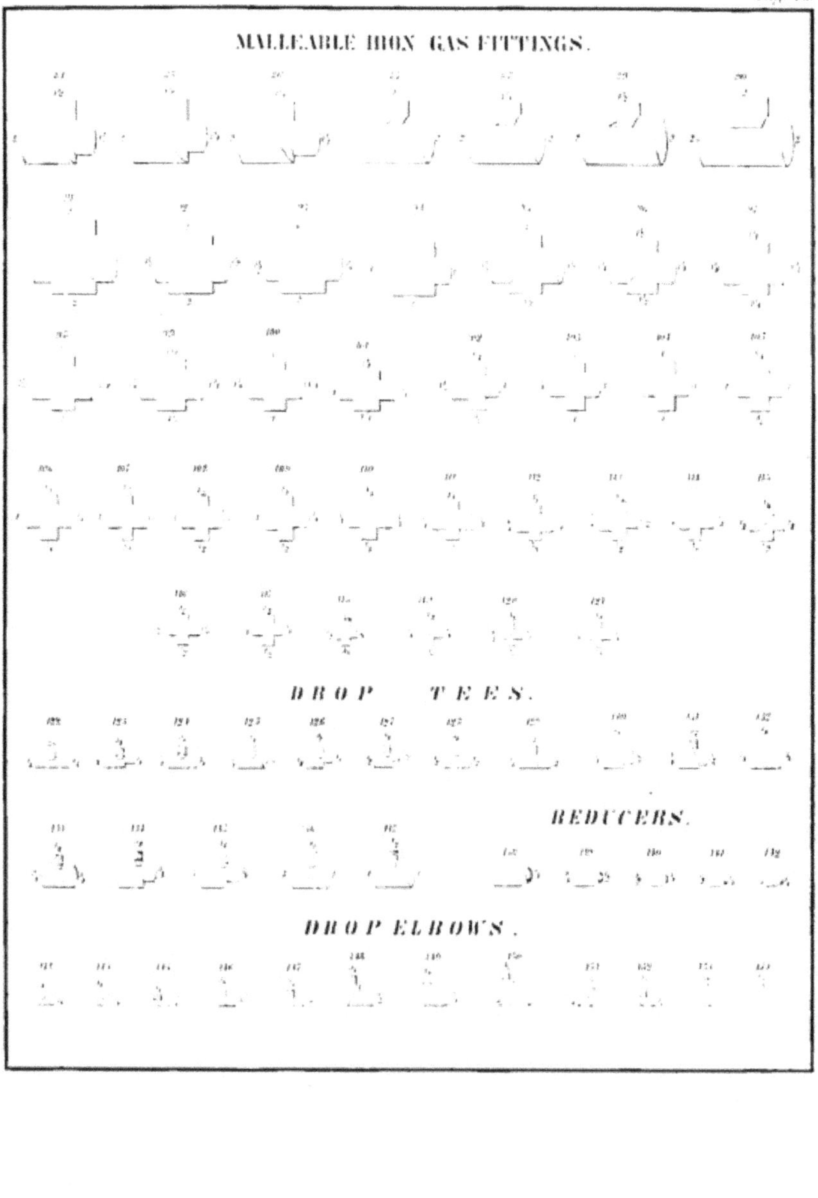
MALLEABLE IRON GAS FITTINGS.
DROP TEES.
REDUCERS.
DROP ELBOWS.

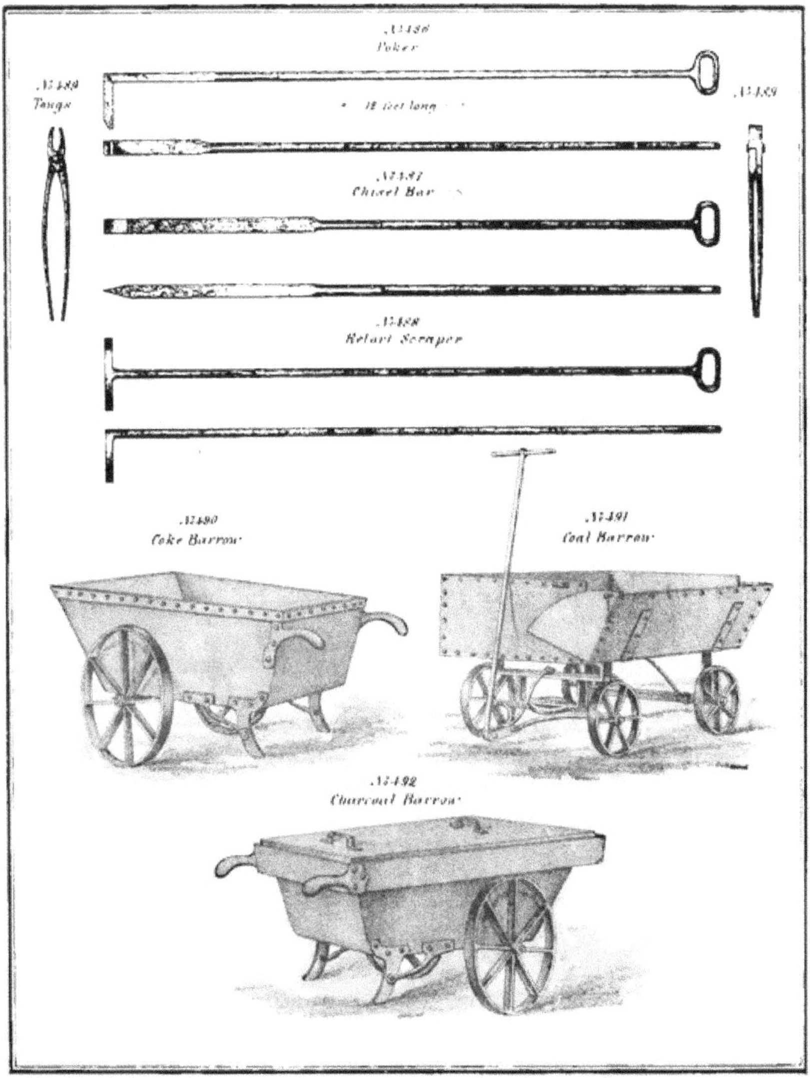
No. 486
Poker
12 feet long
No. 489
Tongs
No. 489
No. 487
Chisel Bar
No. 488
Retort Scraper
No. 490
Coke Barrow
No. 491
Coal Barrow
No. 492
Charcoal Barrow

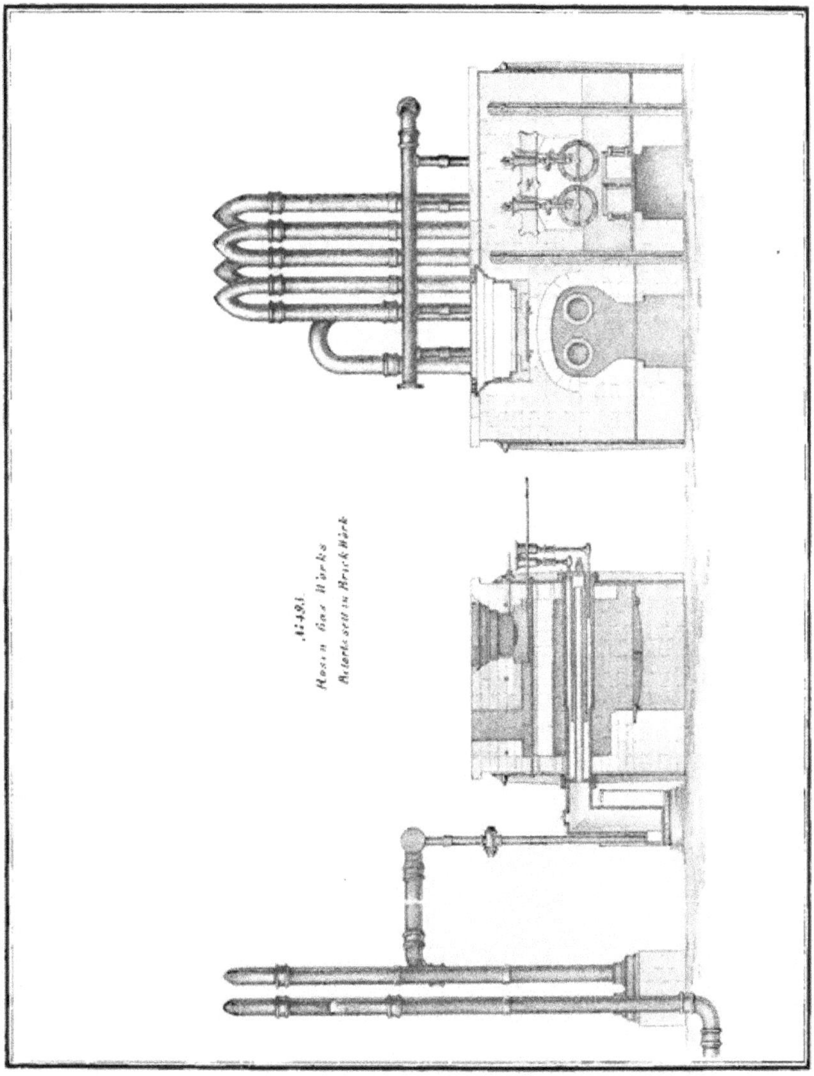

No. 484.
Rosin Gas Works
Retorts set in Brick Work

No. 495

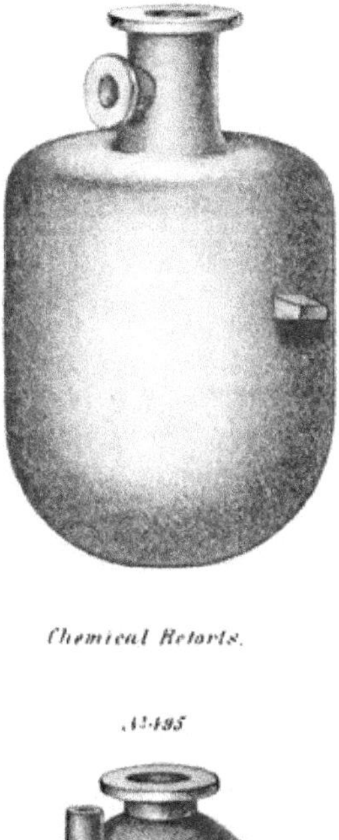

Chemical Retorts.

No. 495

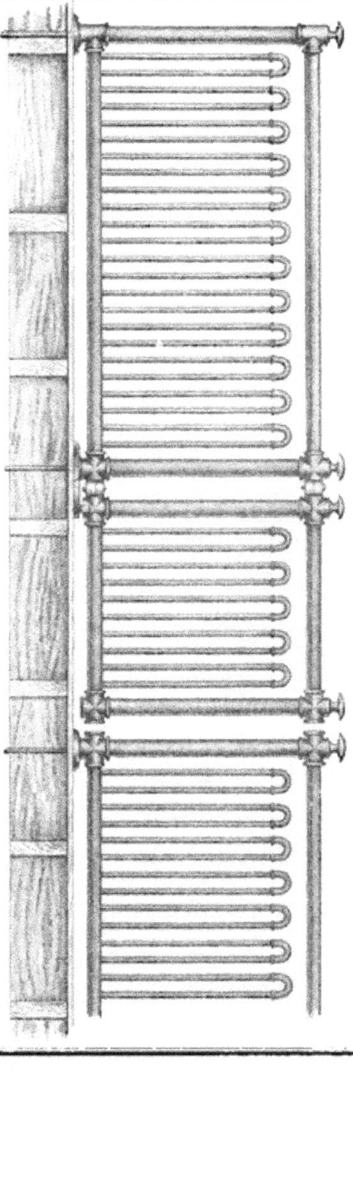

No. 494 Gas Pipe Heating & Steam Radiators Combined

No. 496

Garden Vase

No. 497

Garden Vase

No. 498

Garden Vase

Bolles Rock Drilling Machine

No. 187.

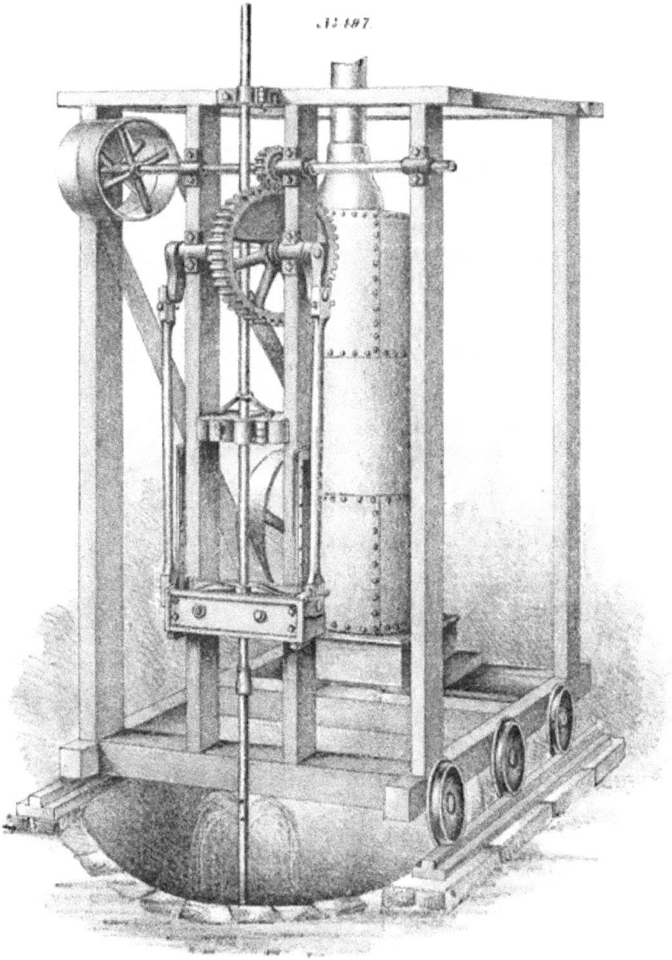

MORRIS, TASKER & Co's ILLUSTRATED CATALOGUE.

Plate 84

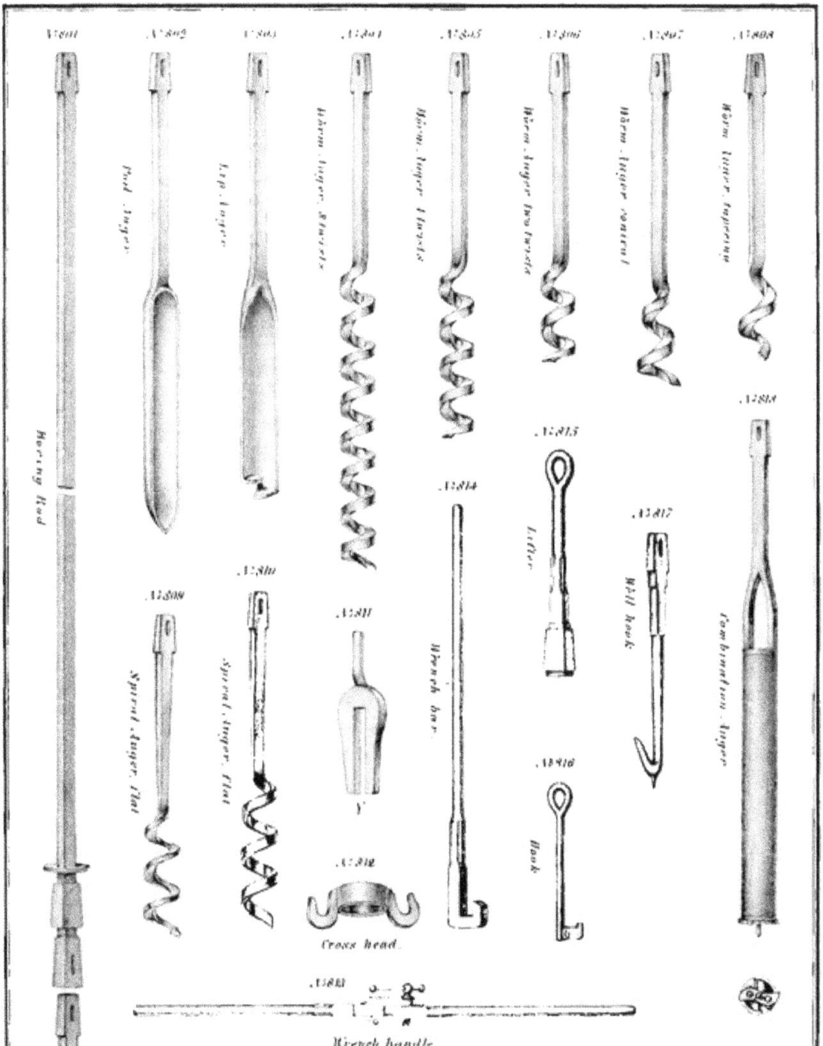

Plate 83

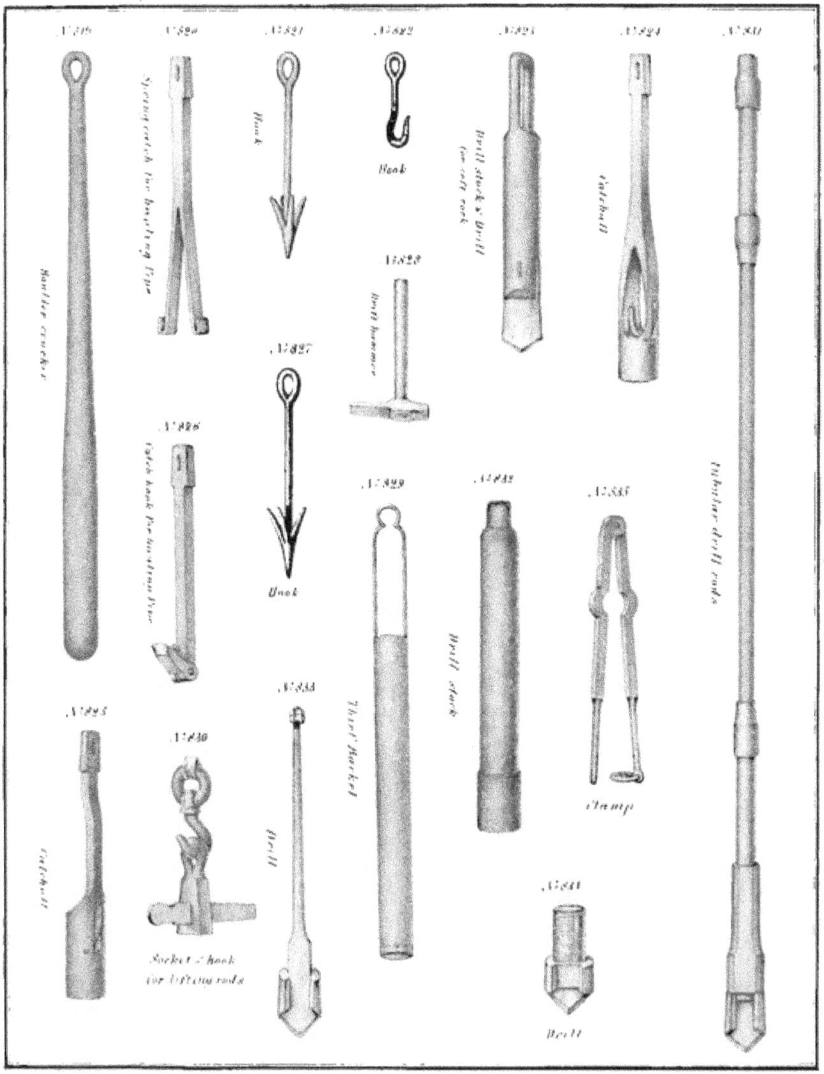
Hook
Hook
Drill hammer
Catchall
Hook
Drill stock
Tubular drill rods
Clamp
Socket & hook for lifting rods
Drill
Drill

Plate 86

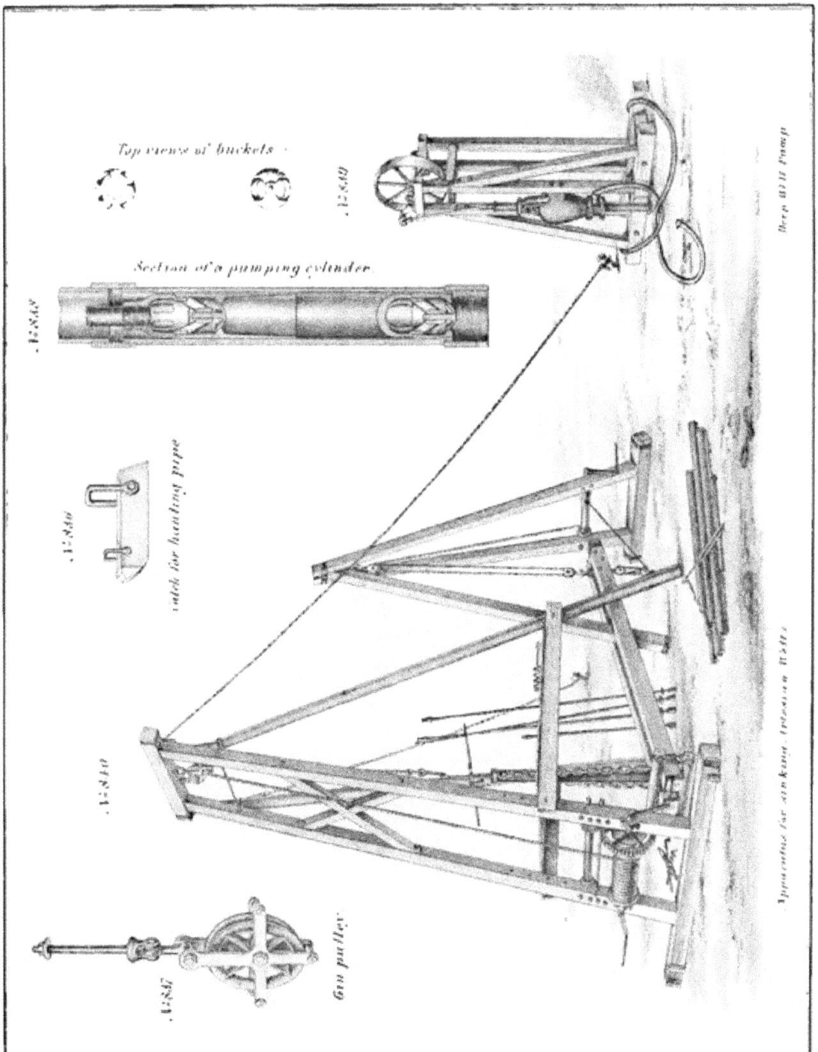
Top views of buckets
Section of a pumping cylinder
catch for hauling pipe
Gin pulley
Deep Well Pump
Apparatus for sinking Artesian Wells

No. 514

Charcoal Barrow

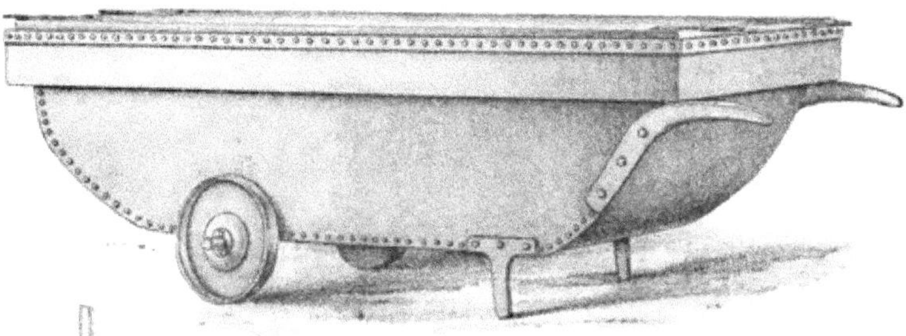

No. 515

Proving Pump for Street Mains

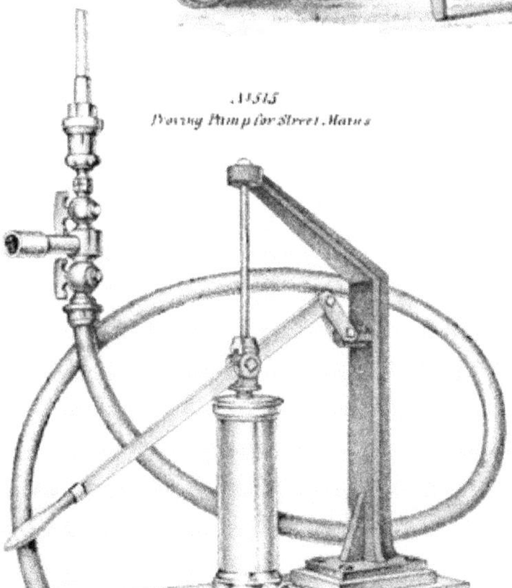

No. 516

Gilmore's Pat. Adjustable Pipe Tong

No. 517

Brown's Pat. Adjustable Pipe Tongs

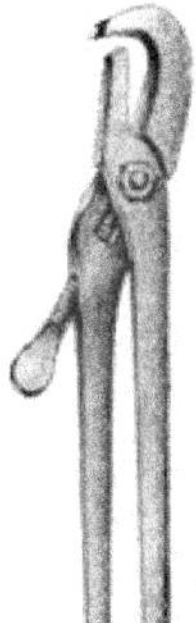

No. 518

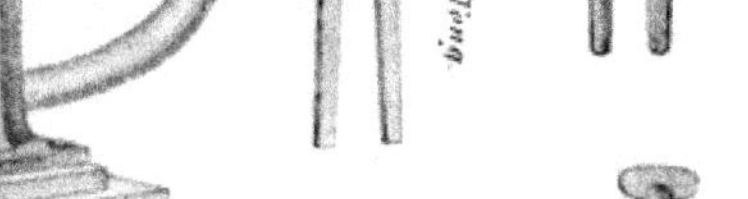

Brown's Pat. Pipe Cutter

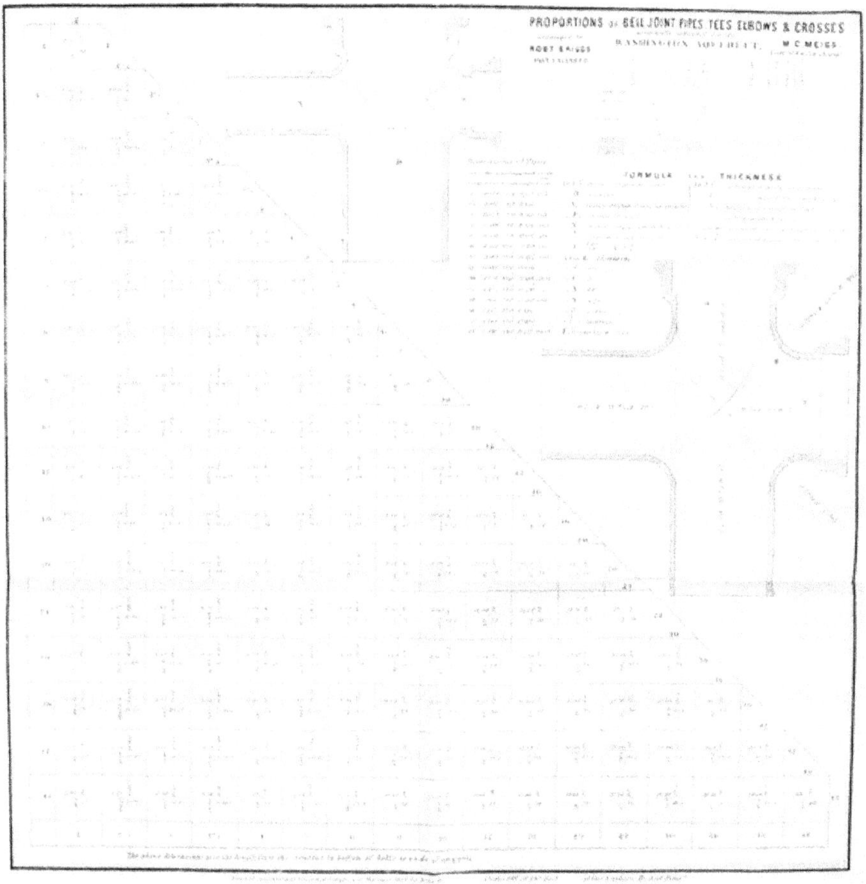
PROPORTIONS OF BELL JOINT PIPES, TEES, ELBOWS & CROSSES
ROBT BRIGGS
M. C. MEIGS
FORMULA
THICKNESS

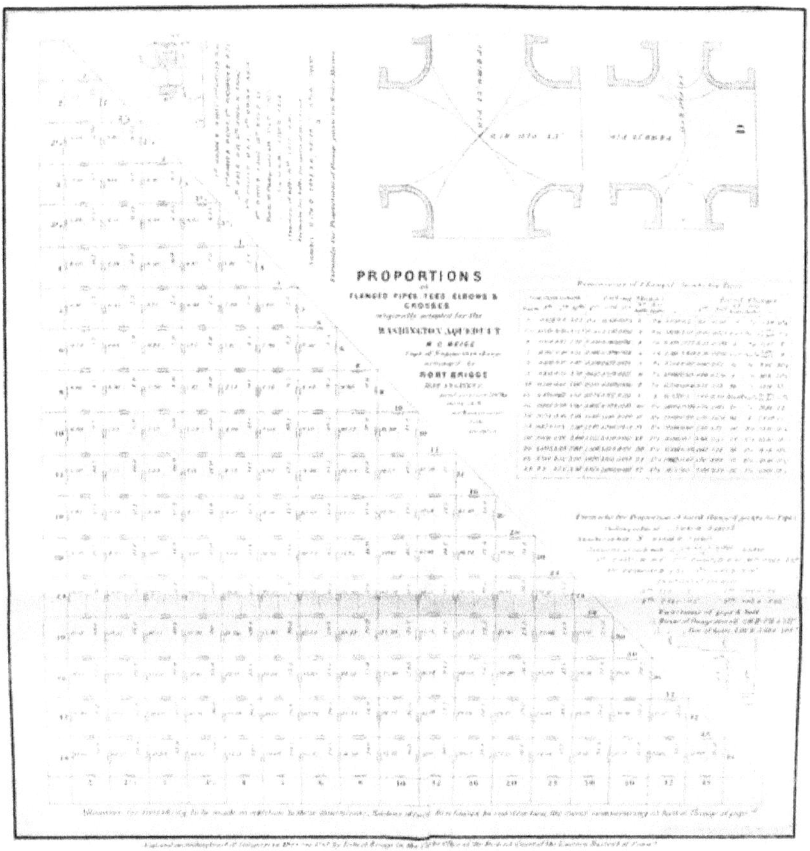
PROPORTIONS
of
FLANGED PIPES TEES ELBOWS &
CROSSES

Deposited 9 May 1871
Morris, Tasker &
Pub. C

MORRIS, TASKER & CO.

[illegible]

Philadelphia

www.ingramcontent.com/pod-product-compliance
Lightning Source LLC
Chambersburg PA
CBHW021659210326
41599CB00013B/1464